模具专业英语教程

（第 2 版）

主　编　叶久新　童长清
副主编　龙艳辉　柳　姣　黄倩文
　　　　刘秀琴　张秀玲
参　编　周玉娥　刘　娟　童　智
　　　　罗正斌　李灶福　张建清

北京理工大学出版社
BEIJING INSTITUTE OF TECHNOLOGY PRESS

内 容 提 要

本书共分 8 章,分别为"模具概论""模具材料简介""冷冲压工艺及模具""塑料成型工艺及模具""锻造工艺及模具""普通机械加工与特种加工""计算机在模具设计与制造中的应用""模具的报价与合同"。内容精炼,选材新颖;所有课文和图例均附有参考译文,旨在提高模具专业学生阅读、翻译模具专业方面的英文资料及参与模具外贸洽谈的能力。

本书可作为高等院校"模具设计与制造"专业以及"材料成型与控制工程"专业的教材,也可作为各类模具技术培训教材,还可供企业从事模具设计、制造以及模具外贸的人员参考。

版权专有　侵权必究

图书在版编目（CIP）数据

模具专业英语教程／叶久新,童长清主编 . — 2 版 . —北京:北京理工大学出版社,2019.8
ISBN 978-7-5682-7487-6

Ⅰ . ①模… Ⅱ . ①叶… ②童… Ⅲ . ①模具-英语-教材 Ⅳ . ①TG76

中国版本图书馆 CIP 数据核字（2019）第 188601 号

出版发行／	北京理工大学出版社有限责任公司
社　　址／	北京市海淀区中关村南大街 5 号
邮　　编／	100081
电　　话／	（010）68914775（总编室）
	（010）82562903（教材售后服务热线）
	（010）68948351（其他图书服务热线）
网　　址／	http://www.bitpress.com.cn
经　　销／	全国各地新华书店
印　　刷／	涿州市新华印刷有限公司
开　　本／	787 毫米×1092 毫米　1/16
印　　张／	16
字　　数／	376 千字
版　　次／	2019 年 8 月第 2 版　2019 年 8 月第 1 次印刷
定　　价／	65.00 元

责任编辑／高　芳
文案编辑／高　芳
责任校对／周瑞红
责任印制／李志强

图书出现印装质量问题,请拨打售后服务热线,本社负责调换

前　言

编者在出版《模具专业英语——设计、报价、结算》及《模具专业英语全程导学》（全译本）后，得到了广大读者的肯定与好评，也不断有读者来电、来信，表达了一书在手便能自学模具英语的愿望。为此，我们在以上两书的基础上编写了《模具专业英语教程》。本书是在上一版的基础上完成的修订版。

本书集《模具专业英语——设计、报价、结算》和《模具专业英语全程导学》之精华，在重视模具设计与制造的基础上，对内容进行了调整和充实，既有关于模具设计与制造方面的最新资料，又带有全程导学的性质，更有利于帮助模具专业学生及模具企业界的外贸人员提高阅读和翻译有关模具外文资料以及直接参与模具外贸洽谈的能力。

本书本着便于学习的目的，按照课文、生词注解、重难点词汇及句子注释、练习、课文导读及参考译文的顺序，对教材内容进行科学合理的编排。书后附有练习参考答案，为模具专业学生及从业人士更好地学习提供了具体而有效的帮助。

本书由叶久新策划、构思，童长清组织、整理，两人共同主编。担任本书副主编的为：龙艳辉、柳姣、黄倩文、刘秀琴、张秀玲；同时参与本书编写的还有周玉娥、刘娟、童智，罗正斌、李灶福、张建清等。

由于编者水平有限，错误之处在所难免，敬请读者批评指正。

编　者

Contents

Chapter 1　Introduction of Mold（模具概论） ··· 1
　Lesson 1　The Definition and Function of Mold ·· 3
　Lesson 2　The Classification of Mold ··· 5

Chapter 2　Introduction of Mold Material（模具材料简介） ··· 9
　Lesson 3　Steels ··· 11
　　Reading Materials（1）　The Function of Alloying Elements in Alloy ······································· 15
　Lesson 4　Heat Treating of Steel ··· 18
　　Reading Materials（2）　Surface Hardening ··· 23

Chapter 3　Press Process and Die Design（冷冲压工艺及模具） ··· 29
　Lesson 5　Forming of Sheet Metals ··· 31
　Lesson 6　Press Process and Product Applications ··· 36
　Lesson 7　Classification of Dies ··· 40
　Lesson 8　Presses ··· 48
　　Reading Materials（3）　Drive Mechanisms for Presses ··· 51
　Lesson 9　Shear Operation ·· 56
　　Reading Materials（4）　Structure of Stamping Die ··· 59
　Lesson 10　Bending Operation ··· 68
　　Reading Materials（5）　An Example of Grouping, Piercing and Bending ······································· 72
　Lesson 11　Drawing Operation ··· 77
　Lesson 12　Compound and Progressive Dies ··· 84
　　Reading Materials（6）　Combination and Compound Dies ··· 87

Chapter 4　Plastics Forming and Mold Design（塑料成型工艺及模具） ··· 93
　Lesson 13　Summary of Plastics ··· 95
　Lesson 14　The Structure of Plastics ··· 99
　　Reading Materials（7）　Additives ··· 102
　Lesson 15　Classification and Application of Plastics ··· 108
　　Reading Materials（8）　Average Plastics ··· 111
　Lesson 16　Injection Molding ··· 117
　　Reading Materials（9）　Classification of Plastics Mold ··· 120
　Lesson 17　Injection Machine ··· 126
　Lesson 18　Representative Structure of Injection Mold ··· 133

Lesson 19　Extrusion Molding ………………………………………………………… 148
　Reading Materials（10） ……………………………………………………………… 150
　　Section A　Compression Molding …………………………………………………… 150
　　Section B　Transfer Molding ………………………………………………………… 151
　　Section C　Blow Molding …………………………………………………………… 152

Chapter 5　Forging Processes and Die Design（锻造工艺及模具）………… 157
Lesson 20　Forging Processes and Die Design ……………………………………… 159

Chapter 6　Basic and Special Machining（普通机械加工与特种加工）……… 165
Lesson 21　Basic Machine Tool Elements …………………………………………… 167
　Reading Materials（11）　Turning …………………………………………………… 172
Lesson 22　Milling ……………………………………………………………………… 181
　Reading Materials（12） ……………………………………………………………… 184
　　Section A　Grinding ………………………………………………………………… 184
　　Section B　Machining Process Selection Factors …………………………………… 187
Lesson 23　Electrical Discharge Machining ………………………………………… 194

Chapter 7　The Application of Computer in Design and Manufacture of Mould and Die（计算机在模具设计与制造中的应用）………………… 199
Lesson 24　Computers and CAD/CAM ……………………………………………… 201
　Reading Materials（13）　CAD/CAM Defined …………………………………… 206

Chapter 8　Quotation and Contract for Mold and Die（模具的报价与合同）……… 215
Lesson 25　Introduction of Quotation for Mold ……………………………………… 217
　Reading Materials（14） ……………………………………………………………… 222
　　Section A　Quotation Strategies and Terms of Payment …………………………… 222
　　Section B　Computerized Price Quoting System for Injection Mold Manufacture …… 226
Lesson 26　Mold Making Contract …………………………………………………… 241

Keys to Exercises …………………………………………………………………… 246

Chapter 1

Introduction of Mold
（模具概论）

Lesson 1

The Definition and Function of Mold

In modern industrial production, mold is an important technical equipment used in shaping process of material (including metal material and nonmetal material). Meanwhile, it is the "magnifying glass of efficiency and profit" to the raw material and equipment, because the value of the final product on die is often tens of, even hundreds of times as valuable as that of the die itself.

Die industry is the basic industry of national economy, and it is called "the mother of industry". Every aspect of human life such as clothes, food, housing and transportation is closely connected with die industry. Therefore, the level of die technology has been a significant symbol to measure a country's developing level of mechanical industry.

Words and Expressions

die [dai]	n.	模型；模具
efficiency [i'fiʃənsi]	n.	效益，功效
magnify ['mægnifai]	v.	放大，扩大
measure ['meʒə]	v.	测量；衡量
mold [məuld]	n.	模具；模型
shape [ʃeip]	v.	使成型；塑造，定型
significant [sig'nifikənt]	adj.	重要的；显著的
symbol ['simbəl]	n.	象征；标志
transportation [trænspɔː'teiʃn]	n.	交通；运输

1. ... because the value of the final product on die is often tens of, even hundreds of times as valuable as that of the die itself.

译文：……因为模具生产出的最终产品的价值，往往是模具自身价值的数十倍，乃至上百倍。

解析：tens of, even hundreds of times：数十倍，乃至上百倍。as + *adj.* + as：和……一样。

2. Every aspect of human life such as clothing, food, housing and transportation is closely connected with die industry.

译文：人类的衣、食、住、行诸方面都离不开模具．

解析：be connected with ：和……有联系/关系。be closely connected with：和……有密切的联系/关系。

Fill in the blanks according to the text.

1. In modern industrial production, _____ is important technical equipment used in _____ process of material (including metal material and nonmetal material).
2. Meanwhile, it is the "_____ glass of efficiency and profit" to the raw material and equipment.
3. _____ is the basic industry of national economy, and it is called "the mother of industry".

（一）课文导读

本课对模具进行了定义，并介绍了模具的重要作用。

（二）课文参考译文

模具的定义和作用

在现代工业生产中，模具是材料（包括金属材料和非金属材料）成型加工的重要工艺装备。同时，它又是原材料及设备的"效益放大器"，因为模具生产出的最终产品的价值，往往是模具自身价值的数十倍，乃至上百倍。

模具工业是国民经济的基础工业，被称为"工业之母"。人类的衣、食、住、行诸方面都离不开模具。所以，模具技术水平的高低已经成为衡量一个国家机械制造业发展水平的重要标志。

Lesson 2

The Classification of Mold

According to the types of raw material, mold can be divided into two main kinds: metal products and nonmetal products.

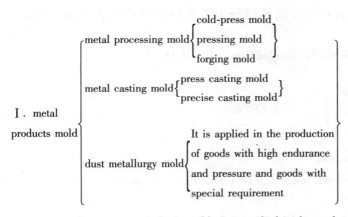

II. nonmetal products mold
- plastic mold: It is applied in the production of plastic goods
- ceramic mold: It is applied in the production of ceramic goods
- rubber mold: It is applied in the production of rubber goods
- glass mold: It is applied in the production of glass goods
- food mold: It is applied in the production of candy and biscuit
- ornament mold: It is applied in the production of ornaments.

 Words and Expressions

ceramics mold	[səˈræmiks]	陶瓷模具
cold-press mold	[kəuldpres]	冷冲模具
food mold	[fuːd]	食品模具
forging mold	[fɔːdʒiŋ]	锻造模具
glass mold	[glɑːs]	玻璃模具
metal product	[ˈmetlˈprɔdʌkt]	金属制品模具

nonmetal product [nɔnˈmetlˈprɔdʌkt]	非金属制品模具
ornament mold [ˈɔːnəmənt]	装饰模具
plastic mold [ˈplæstik]	塑料模具
precise casting mold [priˈsais]	精密铸造模具
press casting mold [presˈkɑːstiŋ]	压力铸造模具
pressing mold [presiŋ]	挤压模具
rubber mold [ˈrʌbə]	橡胶模具

According to the types of raw material, mold can be divided into two main kinds: metal products and nonmetal products.

译文：根据制品所用的原材料种类，模具可分为金属制品模具和非金属制品模具两大类。

解析：according to：依据，按照。be divided into：……被分成……。

Fill in the blanks according to the text.

1. According to the types of raw material, mold can be divided into two main kinds: _____ and _____.
2. Metal products mold can be divided into three kinds: _____, _____ and _____.
3. Metal casting mold can be divided into two kinds: _____ and _____.

（一）课文导读

本课对模具进行了分类，根据制品所用原材料把模具分为金属和非金属模具两大类，并对这两大类进行了细分。

（二）课文参考译文

模具的分类

根据制品所用的原材料种类，模具可分为金属制品模具和非金属制品模具两大类：

$$\text{Ⅰ. 金属制品模具} \begin{cases} \text{金属加工模具} \begin{cases} \text{冷冲模具} \\ \text{挤压模具} \\ \text{锻造模具} \end{cases} \\ \text{金属铸造模具} \begin{cases} \text{压力铸造模具} \\ \text{精密铸造模具} \end{cases} \\ \text{粉末冶金模具：用于生产耐磨、耐压及特定要求的产品} \end{cases}$$

这些模具广泛用于机电产品、汽车、航空仪器及其他金属制品

$$\text{Ⅱ. 非金属制品模具} \begin{cases} \text{塑料模具：用于各种塑料制品的生产} \\ \text{陶瓷模具：用于各种陶瓷制品的生产} \\ \text{橡胶模具：用于各种橡胶制品的生产} \\ \text{玻璃模具：用于各种玻璃制品的生产} \\ \text{食品模具：用于各种糖果、饼干的生产} \\ \text{装饰模具：用于各种装饰制品的生产} \end{cases}$$

Chapter 2

Introduction of Mold Material
（模具材料简介）

Lesson 3

Steels

Steels (first made in China and Japan around 600—800 AC) are generally divided into the categories of carbon steels and alloy steels (including tool steels).

Carbon Steels

Carbon steels are used extensively in tool construction. Carbon steels are those steels which only contain iron and carbon, and small amounts of other alloying elements. Carbon steels are the most common and least expensive type of steels used for tools. The three principal types of carbon steels used for tooling are low carbon, medium carbon, and high carbon steels. Low carbon steel contains between 0.05% and 0.3% carbon. Medium carbon steel contains between 0.3% and 0.7% carbon. And high carbon steel contains between 0.7% and 1.5% carbon. As the carbon content is increased in carbon steel, the strength, toughness, and hardness also increase when the metal is heat treated.

Low carbon steels are soft, tough steels that are easily machined and welded. Due to their low carbon content, these steels cannot be hardened except by case hardening. Low carbon steels are well suited for the following applications: tool bodies, handles, die shoes, and similar situations where strength and wear resistance are not required.

Medium carbon steels are used where greater strength and toughness are required. Since medium carbon steels have a higher carbon content they can be heat treated to make parts such as studs, pins, axles, and nuts. Steels in this group are more expensive as well as more difficult to machine and weld than low carbon steels.

High carbon steels are the most hardenable type of carbon steel and are used frequently for parts where wear resistance is an important factor. Other applications where high carbon steels are well suited include drill bushings, locators, and wear pads. Since the carbon content of these steels is so high, parts made from high carbon steel are normally difficult to machine and weld.

Alloy Steels

Alloy steels are basically carbon steels with additional elements added to alter the characteristics and bring about a predictable change in the mechanical properties of the alloyed metal. Alloy steels are not normally used for most tools due to their increased cost, but some have found favor for special applications. The alloying elements used most often in steels are manganese, nickel, molybdenum, and chromium.

Another type of alloy steel frequently used for tooling applications is stainless steel. Stainless steel is a term used to describe high chromium and nickel-chromium steels. These steels are used for tools which must resist high temperatures and corrosive atmospheres. Some high chromium steels can be hardened by heat treatment and are used where resistance to wear, abrasion, and corrosion are required. Typical applications where a hardenable stainless steel is sometimes preferred are plastic injection molds: Here the high chromium content allows the steel to be highly polished and prevents deterioration of the cavity from heat and corrosion.

Words and Expressions

abrasion [əˈbreiʒən]	n.	磨损
alloy steel		合金钢
alter [ˈɔːltə]	v.	改变
axle [ˈæksl]	n.	轴，车轴，轮轴
carbon steel		碳素钢
category [ˈkætəɡəri]	n.	类别，种类，范畴
chromium [ˈkrəumjəm]	n.	铬
corrosive [kəˈrəusiv]	adj.	腐蚀的，蚀坏的；腐蚀性的
	n.	腐蚀物；腐蚀剂
deterioration [diˌtiəriəˈreiʃən]	n.	变坏；退化；堕落
die shoe		模脚
drill bushing		钻套
element [ˈelimənt]	n.	要素，元素，成分，元件
frequently [ˈfriːkwəntli]	adv.	常常，频繁地，经常地
generally [ˈdʒenərəli]	adv.	一般，通常，一般地
handle [ˈhændl]	n.	柄，把手；把柄
	vt.	触摸，运用，买卖，处理，操作
hardenable	adj.	可硬化的
heat treat		热处理
impart [imˈpɑːt]	vt.	给予(尤指抽象事物)；传授；告知；透露
iron [ˈaiən]	n.	铁；熨斗；坚强；烙铁；镣铐
locator [ləuˈkeitə]	n.	定位器(表示位置之物)；土地
machine [məˈʃiːn]	n.	机器，机械
	vt.	用机器制造，用车床加工
manganese [ˌmæŋɡəˈniːz, ˈmæŋɡəniːz]	n.	[化]锰(元素符号为 Mn)
molybdenum [məˈlibdinəm]	n.	[化]钼

Chapter 2　Introduction of Mold Material

nickel ['nikl]	n.	[化]镍，镍币
nut [nʌt]	n.	螺母，螺帽
pin [pin]	n.	钉；销；栓
polish ['pɔliʃ]	n.	磨光；光泽
	vt.	擦亮；发亮；磨光
predictable [pri'diktəb(ə)l]	adj.	可预言的
preferred [pri'fə:d]	adj.	首选的
principal ['prinsəp(ə)l,-sip-]	adj.	主要的，首要的
resist [ri'zist]	vt.	抵抗，反抗，抗，忍得住
stud [stʌd]	n.	圆柱头螺栓
wear pad		耐磨垫板
wear resistance		耐磨性
weld [weld]	vt.	焊接
	n.	焊接；焊缝

1. Carbon steels are those steels which only contain iron and carbon, and small amounts of other alloying elements. Carbon steels are the most common and least expensive type of steels used for tools.

译文：碳钢是指那些仅仅由铁和碳以及少量的其他合金元素构成的钢。碳钢是一种最常见的、最廉价的制造工具的钢。

解析：此句中 which 是关系代词，引导定语从句，修饰前面的名词 steels，并充当定语从句的主语。used for tools 是过去分词短语作后置定语，修饰中心词 steels，它也可等同于定语从句 which/that are used for tools。

2. Low carbon steels are well suited for the following applications: tool bodies, handles, die shoes, and similar situations where strength and wear resistance are not required.

译文：低碳钢很适合于以下用途：工具主体、把柄、模脚以及一些不要求强度和耐磨性的类似情况。

解析：此句中 where 是关系副词，引导定语从句，修饰前面的名词 similar situations。

3. Since medium carbon steels have a higher carbon content they can be heat treated to make parts such as studs, pins, axles, and nuts.

译文：由于中碳钢含碳量较高，所以经过热处理后可被制成如圆柱头螺栓、销、轴和螺母之类的部件。

解析：此句中 since 是从属连词，引导原因状语从句，译为"由于"。句中 studs, pins, axles, and nuts 译为"圆柱头螺栓、销、轴和螺母"。

4. High carbon steels are the most hardenable type of carbon steel and are used frequently for parts where wear resistance is an important factor.

译文：高碳钢是一种最可硬化的碳钢，常用于对耐磨性要求非常高的部件。

解析：此句中 where 是关系副词，引导定语从句。

5. The alloying elements used most often in steels are manganese, nickel, molybdenum, and chromium.

译文：钢中最常用的合金化元素是锰、镍、钼和铬。

解析：此句中 element 译为"元素"。它还有"适合的环境"之意。例如：

You're in your element once the conversation turns to movies.

谈起电影，你便如鱼得水。

6. These steels are used for tools which must resist high temperatures and corrosive atmospheres.

译文：这些钢用于必须耐高温和耐腐蚀性环境的工具。

解析：此句中 resist 译为"耐……"，resist doing 译为"防……"。

7. Some high chromium steels can be hardened by heat treatment and are used where resistance to wear, abrasion, and corrosion are required.

译文：一些高铬钢能够被热处理硬化，用于有耐磨损、耐大气腐蚀要求的场合。

解析：此句中 where 是关系副词，引导状语从句。

8. Typical applications where a hardenable stainless steel is sometimes preferred are plastic injection molds: Here the high chromium content allows the steel to be highly polished and prevents deterioration of the cavity from heat and corrosion.

译文：常选用可硬化不锈钢的典型场合是塑料注射模具：因为高含铬量使得钢材具有很好的抛光性，而且能够防止型腔由于受热和腐蚀而产生性能的退化。

解析：此句中 prevent...from 译为"阻止，妨碍"。

Exercise

Fill in the blanks according to the text.

1. The three principal types of carbon steels used for tooling are low carbon, _____ carbon, and high carbon steels.
2. Low carbon steels are _____, tough steels that are easily machined and welded.
3. Since medium carbon steels have a higher carbon content they can be heat treated to make parts _____ studs, pins, axles, and nuts.
4. _____ the carbon content of these steels is so high, parts made from high carbon steel are normally difficult to machine and weld.
5. Alloy steels are not normally used for most tools _____ their increased cost, but some have found favor for special applications.

Chapter 2　Introduction of Mold Material

 (1)

The Function of Alloying Elements in Alloy

Various alloying elements are added to iron in order to impart certain properties. These are summarized below. The major detrimental effects are stated in parentheses.

Carbon: hardenability, strength, hardness, and wear resistance.

Nickel: strength and toughness; minor effect on hardenability.

Chromium: strength, toughness, hardness, and wear resistance; increases depth of hardness penetration in heat treatment.

Molybdenum: hardenability, wear resistance, toughness; strength, creep resistance, and hardness at elevated temperatures.

Vanadium: strength, abrasion resistance, hardness at elevated temperatures; inhibits grain growth during heat treatment.

Copper: resistances to atmosphere corrosion, improve strength with little loss in ductility. (Can adversely affect surface quality and hot-working characteristics.)

Manganese: hardenability and ductility.

Lead: machinability. (Causes liquid-metal embrittlement.)

Sulfur: machinability. (Lowers impact strength and transverse ductility; impairs surface quality and weldability.)

Silicon: strength, high electrical conductivity; decreases magnetic hysteresis loss.

Phosphorus: strength, hardenability, corrosion resistance, machinability. (Decreases ductility and toughness.)

Boron: hardenability.

Tungsten and Cobalt: strength and hardness at elevated temperatures.

Columbium (Niobium): fine grain size, strength, lowers transition temperature.

Tellurium: machinability of leaded steels.

Zirconium and Cerium: control shape of inclusions (sulfides) and improve toughness in high-strength, low-alloy steels.

Aluminum, Silicon and Calcium: added to steels during solidification to remove oxygen and nitrogen.

 Words and Expressions

aluminum　　　　　　　　　　　　　　　　n.　[化]铝

cerium	n.	[化]铈
cobalt	n.	[化]钴(符号为 Co)；钴类颜料；由钴制的深蓝色
columbium	n.	[化]钶
detrimental	adj.	有害的
ductility	n.	展延性；柔软性；顺从
hardenability	n.	[冶]可硬性，淬透性，可淬性
hysteresis	n.	滞后作用；[物]磁滞现象
niobium	n.	[化]铌
phosphorus	n.	磷
solidification	n.	凝固
sulfur	n.	[化]硫磺
	vt.	用硫磺处理
tellurium	n.	碲
transverse	adj.	横向的；横断的
tungsten	n.	[化]钨
vanadium	n.	[矿]钒
zirconium	n.	锆

课文导读及参考译文

（一）课文导读

本文主要介绍钢的特点和用途。钢通常被分为碳钢和合金钢（包括工具钢），其中用于工具制造的碳钢分为三种类型：低碳钢、中碳钢和高碳钢。

（二）课文参考译文

钢

钢（公元前 600—800 年左右首先在中国和日本制造出来）通常被分为碳钢和合金钢（包括工具钢）两大类。

碳钢

碳钢广泛用于制造工具。碳钢是指那些仅仅由铁和碳以及少量的其他合金元素构成的钢。碳钢是一种最常见的、最廉价的制造工具的钢。用于制造工具的碳钢分为三种主要类型：低碳钢、中碳钢和高碳钢。低碳钢含碳量为 0.05%～0.3%。中碳钢含碳量为 0.3%～0.7%。高碳钢含碳量为 0.7%～1.5%。经过热处理的碳钢的强度、韧性和硬度随着含碳量的增加而增大。

低碳钢是易被机加工和焊接的软、韧钢。由于含碳量低，除了通过表面硬化处理，这些钢不能被硬化。低碳钢很适合于以下用途：工具主体、把柄、模脚以及一些不要求强度和耐磨性的类似情况。

中碳钢应用于对强度和韧性要求较高的部件。由于中碳钢含碳量较高，所以经过热处理后可被制成如圆柱头螺栓、销、轴和螺母之类的部件。中碳钢比低碳钢贵，也比低碳钢难于机加工和焊接。

高碳钢是最可硬化的一种碳钢，常用于对耐磨性要求非常高的部件。适合用高碳钢制作的其他部件包括衬套、定位销和耐磨垫板。由于这些钢的碳含量很高，所以用高碳钢制成的部件通常难以机加工和焊接。

合金钢

合金钢本质上是含有其他元素的碳钢，添加其他元素的目的是为了改变碳钢的特征，并使其力学性能发生可以预见的变化。合金钢成本高，因此不能广泛应用于大多数工具，但是一些合金钢适合于特殊用途。钢中最常用的合金化元素是锰、镍、钼和铬。

经常用于制造工具的另一种合金钢是不锈钢。不锈钢是指高铬钢和高镍-铬钢。这些钢应用于必须耐高温和耐腐蚀性空气的部件。一些高铬钢能够通过热处理硬化，应用于要求耐磨损、耐腐蚀的场合。常选用可硬化不锈钢的典型场合是塑料注射模具：因为高含铬量使得钢材具有很高的抛光性，而且能够防止型腔由于受热和腐蚀而产生性能的退化。

（三）阅读材料（1）参考译文

合金中合金化元素的作用

为了获得某些性能，不同的合金化元素被加入铁中。合金化元素对性能的影响概括如下，主要的有害影响列于括号中。

碳：可硬性、强度、硬度和耐磨性。

镍：强度和韧度；对可硬性影响较小。

铬：强度、韧性、硬度和耐磨性；增加热处理时的淬硬深度。

钼：可硬性、耐磨性、韧性、强度、抗蠕变性和高温硬度。

钒：强度、耐磨性、高温硬度；热处理过程中抑制晶粒生长。

铜：抗大气腐蚀性、在不降低延展性的条件下提高强度。（对表面质量和热工作特性有不利影响。）

锰：可硬性和延展性。

铅：切削加工性。（引起液体-金属脆化。）

硫：切削加工性。（降低冲击强度和横向延展性；降低表面质量和可焊性。）

硅：强度、高导电性；减少磁滞损耗。

磷：强度、可硬性、耐蚀性、切削加工性。（降低延展性和韧性。）

硼：可硬性。

钨和钴：高温强度和硬度。

铌：细化晶粒、强度、降低转变温度。

碲：含铅钢的切削加工性。

锆和铈：控制杂质（硫化物）形状，改善高强低合金钢的韧性。

铝、硅和钙：钢水凝固时加入钢中以除去氧和氮。

Lesson 4

Heat Treating of Steel

Specifications for heat-treating processes are among the most important of those shown on an engineering drawing. Proper heat treatment is a powerful tool for developing the best possible properties that a material can possess. In general, heat treatment may be described as a combination of heating and cooling operations, timed and applied to a metal or an alloy in the solid state in a way that will produce desired properties. Principally, heat treatment is used to produce strengthening, but some heat-treating processes soften, toughen, or otherwise enhance properties.

Internally, a metal or an alloy consists of one or more kinds of atoms packed together in orderly three-dimensional arrangements called crystals. The crystals, in turn, are bonded together in diverse ways which are described in terms of microstructure or grain structure. Any given structure can be altered to some extent by plastic deformation from compressive, tensile, or shear forces, but the available time-temperature treatments provide a greater variety of properties. Heat treatments are carefully controlled combinations of such variables as time, temperature, rate of temperature change, and furnace atmosphere. The selection of a specific treatment must be based upon knowledge of the properties desired in the finished part.

There is available today a multitude of metals and alloys designed for various purposes. There are also many different heat-treating processes. Not all the treatments can be used with each metal or alloy. In other words, the treatment selected must be one that is compatible with the specified material. Heat treatment cannot be selected independently of material. One is just as important as the other.

Reasons for Heat Treating

Ferrous Metals. Ferrous parts are heat-treated for several reasons: to relieve internal stresses, to change the microstructure by refining the grain size or producing uniform grain throughout a part, to alter the surface chemistry by adding or deleting elements, and to strengthen a metal part.

Treatment of Ferrous Materials. Iron is the major constituent in the steels used in tooling, to which carbon is added in order that the steel may harden. Alloys are put into steel to enable it to develop properties not possessed by plain carbon steel, such as the ability to harden in oil or air, increased wear resistance, higher toughness, and greater safety in hardening.

Heat treatment of ferrous materials involves several important operations which are customarily referred to under various headings, such as normalizing, spheroidizing, stress relieving, annealing, hardening, tempering, and case hardening.

Normalizing involves heating the material to a temperature of about 100 °F – 200 °F (55 ℃ – 100 ℃) above the critical range and cooling in still air. This is about 100 °F (55 ℃) over the regular hardening temperature.

The purpose of normalizing is usually to refine grain structures that have been coarsened in forging. With most of the medium-carbon forging steels, alloyed and unalloyed, normalizing is highly recommended after forging and before machining to produce more homogeneous structures, and in most cases, improved machinability.

High-alloy air-hardened steels are never normalized, since to do so would cause them to harden and defeat the primary purpose.

Spheroidizing is a form of annealing which, in the process of heating and cooling steel, produces a rounded or globular form of carbide — the hard constituent in steel.

Tool steels are normally spheroidized to improve machinability. This is accomplished by heating to a temperature to 1 380 °F – 1 400 °F (749 ℃ – 760 ℃) for carbon steels and higher for many alloy tool steels, holding at heat one to four hours, and cooling slowly in the furnace.

Stress Relieving. This is a method of relieving the internal stresses set up in steel during forming, cold working, and cooling after welding or machining, it is the simplest heat treatment and is accomplished merely by heating to 1 200 °F – 1 350 °F (649 ℃ – 732 ℃) followed by air or furnace cooling.

Large dies are usually roughed out, then stress-relieved and finish-machined. This will minimize change of shape not only during machining but during subsequent heat treating as well. Welded sections will also have locked-in stresses owing to a combination of differential heating and cooling cycles as well as to changes in cross section. Such stresses will cause considerable movement in machining operations.

Annealing. The process of annealing consists of heating the steel to an elevated temperature for a definite period of time and, usually, cooling it slowly. Annealing is done to produce homogenization and to establish normal equilibrium conditions, with corresponding characteristic properties.

Tool steel is generally purchased in the annealed condition. Sometimes it is necessary to rework a tool that has been hardened, and the tool must then be annealed. For this type of anneal, the steel is heated slightly above its critical range and then cooled very slowly.

Hardening. This is the process of heating to a temperature above the critical range, and cooling rapidly enough through the critical range to appreciably harden the steel.

Tempering. This is the process of heating quenched and hardened steels and alloys to some temperature below the lower critical temperature to reduce internal stresses setup in hardening.

Case Hardening. The addition of Carbon to the surface of steel parts and the subsequent hardening operations are important phases in heat treating. The process may involve the use of molten sodium cyanide mixtures, pack carburizing with activated solid material such as charcoal or coke, gas or oil carburizing, and dry cyaniding.

Words and Expressions

accomplished [ə'kɔmpliʃt]	adj.	完成的；熟练的；多才多艺的
air-hardened	adj.	空气(冷却)硬(化)的，自硬的
annealing [æ'ni:liŋ]	v.	退火
be compatible with	v.	适合；一致
carburizing ['kɑ:bjuraiziŋ]	n.	增碳剂，渗碳剂
case hardening		表面硬化
charcoal ['tʃɑ:kəul]	n.	木炭
coke [kəuk]	n.	可乐；焦炭
	v.	(使)成焦炭
critical temperature	n.	临界温度
cyaniding ['saiənaidiŋ]	n.	氰化
elevated ['eliveitid]	adj.	提高的；严肃的；欢欣的
equilibrium [ˌi:kwi'libriəm]	n.	平衡；均衡；保持平衡的能力；沉着，平静，安静
ferrous ['ferəs]	adj.	铁的，含铁的；[化]亚铁的
finish machined	n.	精加工
furnace ['fə:nis]	n.	炉子，熔炉
furnace cooling		(随)炉冷却
grain size		粒度，颗粒尺寸；结晶粒度[大小]
hardening ['hɑ:dəniŋ]	v.	淬火
homogeneous [ˌhɔməu'dʒi:njəs]	adj.	同类的，相似的；均一的，均匀的
homogenization [ˌhəumədʒənai'zeiʃən]	n.	(均)匀化，均质化；同质化，纯一化
independently [indipendəntli]	adv.	独立地；自立地
locked-in ['lɔkt'in]	adj.	牢固的
machinability [məʃi:nə'biliti]	n.	机械加工性切削性
microstructure ['maikrəu'strʌktʃə]	n	微观结构，显微结构
molten sodium cyanide		熔融的氰化钠
multitude ['mʌltitju:d]	n.	多数；群众
normalizing ['nɔ:məlaiziŋ]		常化，正火
refining [ri'fainiŋ]	n.	精炼
shear [ʃiə]	v.	剪；修剪；剪切
specify ['spesifai]	vt.	指定，详细说明，列入清单
spheroidizing ['sfiərɔidaiziŋ]		球化(处理)
stress relieving		应力消除
subsequent ['sʌbsikwənt]	adj.	后来的；并发的

tempering ['tempəriŋ]	v.	回火
tensile ['tensail]	adj.	可拉长的,可伸长的;[物]张力的,拉力的
uniform ['ju:nifɔ:m]	adj.	统一的,相同的,一致的
	n.	制服
	vt.	使……成一样;使……穿制服
welded section		焊接区域

1. In general, heat treatment may be described as a combination of heating and cooling operations, timed and applied to a metal or an alloy in the solid state in a way that will produce desired properties.

译文:通常,热处理被描述为对金属或合金按一定的速度进行加热和冷却的一种处理过程,以此来获得理想的性能。

解析:describe as:描述为。apply to:将……应用于……。in a way 译为"在某种程度上,稍稍"。又如:

I like the new styles, in a way. 在某种程度上我喜欢这些新款式。

2. Any given structure can be altered to some extent by plastic deformation from compressive, tensile, or shear forces, but the available time-temperature treatments provide a greater variety of properties.

译文:由压缩、拉伸或剪切等作用力带来的塑性变形不但可以在某种程度上改变任何给定的物质结构,而且还能结合给定的热处理工艺获得较宽范围内的性能的变化。

解析:句中 given 译为"特定的,给定的"。又如:

We will meet at a given time and location. 我们将在指定的时间和地点见面。

3. The selection of a specific treatment must be based upon knowledge of the properties desired in the finished part.

译文:具体热处理工艺的选择必须以成品零件所期望的性能为基础。

解析:此句中 based upon 译为"以……为基础"。又如:

Your conclusion should be based upon careful research. 你应该以审慎的研究为基础而下结论。

4. Not all the treatments can be used with each metal or alloy. In other words, the treatment selected must be one that is compatible with the specified material.

译文:并非所有的热处理都能用于每种金属或合金。换句话说,选择的热处理工艺必须和指定材料所需要的热处理工艺相一致。

解析:句中 not all 表部分否定,有时相当于 all...not...。又如:

All is not gold that glitters. 发光的并不总是金子。

5. Heat treatment of ferrous materials involves several important operations which are customarily referred to under various headings, such as normalizing, spheroidizing, stress relieving, annealing, hardening, tempering, and case hardening.

译文：含铁材料的热处理包含几个重要操作，常常被冠以如下称谓，例如正火、球化、应力消除、退火、淬火、回火和表面硬化。

解析：句中 such as 译为"例如……"，"像这种的"。refer to：提到，谈到，指的是。

6. The purpose of normalizing is usually to refine grain structures that have been coarsened in forging.

译文：正火的目的通常是为了细化锻造过程中被粗化的晶粒。

解析：句中 grain structures 译为"晶粒"。that 在句中是关系代词，引导定语从句，修饰先行词 grain structures，并代替先行词在从句中作主语。

7. Large dies are usually roughed out, then stress-relieved and finish-machined.

译文：大型模具通常首先制坯，然后消除应力和精加工。

解析：此句中 die 译为"模具"（名词）。

8. Welded sections will also have locked-in stresses owing to a combination of differential heating and cooling cycles as well as to changes in cross section.

译文：不同加热和冷却循环的组合以及横截面的改变，使得焊接区域也存在牢固的应力。

解析：句中 locked-in 是形容词，译为"牢固的"。owe to：归功于，由于。as well as 译为"也，又"。cross section 译为"横截面"。

9. Sometimes it is necessary to rework a tool that has been hardened, and the tool must then be annealed.

译文：有时有必要再次加工已经硬化的工件，那么这个工件必须退火。

解析：此句中 that 是关系代词，引导定语从句，修饰先行词 tool，并代替先行词在从句中作主语。句中 it is necessary to 译为"有必要……"。又如：

Is it necessary for you to buy so many dresses at a time?

你有必要一次买那么多衣服吗？

10. The process of annealing consists of heating the steel to an elevated temperature for a definite period of time and, usually, cooling it slowly. Annealing is done to produce homogenization and to establish normal equilibrium conditions, with corresponding characteristic properties.

译文：退火过程常为加热钢至高温并保持一定的时间，然后慢慢冷却。退火是为了达到均一化并且建立正常的平衡条件，以获取相应的性能。

解析：句中 consist of 译为"由……组成"。

11. The process may involve the use of molten sodium cyanide mixtures, pack carburizing with activated solid material such as charcoal or coke, gas or oil carburizing, and dry cyaniding.

译文：这个过程包括使用熔融氰化钠混合物，用活性固体材料例如木炭、焦炭或者用能够提供碳原子的气体、油类和干的氰化物填满渗碳介质。

解析：此句中 such as 译为"例如"。

Chapter 2　Introduction of Mold Material

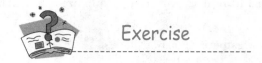

Exercise

Fill in the blanks according to the text.

1. Proper heat treatment is a powerful tool for developing the best possible properties _____ a material can possess.
2. Internally, a metal or an alloy consists _____ one or more kinds of atoms packed together in orderly three-dimensional arrangements called crystals.
3. Iron is the major _____ in the steels used in tooling, to which carbon is added in order that the steel may harden.
4. This will minimize change of shape not only during machining _____ during subsequent heat treating as well.
5. The addition of Carbon to the surface of steel parts and the _____ hardening operations are important phases in heat treating.

Reading Materials (2)

Surface Hardening

Case hardening, differential hardening, or surface hardening produces a hard skin around the core of a ferrous part. Several processes can be used for case hardening. All but one, induction hardening, involve diffusion of another material into the metal surface.

Carburizing: The most common diffusion-type hardening process is carburizing. Briefly, the carbon diffuses into the metal at a high temperature (about 1 700 °F, or 927 °C) and actually transforms the surface layer of the metal from a low-carbon steel to a high-carbon steel. Then, a simple quench and temper hardens the high-carbon skin (because hardness is a function of carbon content). The hardened case resists wear and scuffing, and provides higher fatigue resistance because of the beneficial compressive stresses at the surface. Tensile stresses at the surface are harmful to fatigue performance.

Specifications for carburizing must include core as well as case requirements. Closely controlled case properties usually require gas carburizing because the carbonaceous atmosphere can be monitored closely. The carbon atmosphere must completely surround the part for uniform case depth and hardness. Of course, carbon content gradually decreases from the surface of the case to the core, and so does hardness. The distinction between actual core and case, or "effective case", is sometimes defined as the point where hardness measures Rockwell C50. As little metal should be removed from the surface as possible after carburizing to prevent removing the best part of the hardened case: generally, no more than 20% of the carburized case should be removed after heat treatment.

Typical specifications for carburizing may include:

1. Case hardness on surface.

2. Effective case depth.

3. Carbide condition and restriction.

4. Hardness gradient across case.

5. Core hardness.

6. Core structure (fully martensitic, no austenite, etc).

7. NDT (testing) for surface cracking.

8. Maximum stock removal after carburizing.

Variations on the basic carburizing cycle are designed to fit special needs:

Carbonitriding: Carbonitriding sounds like a combination of carburizing and nitriding. Actually, carbonitriding is essentially liquid carburizing, but it also involves some diffusion of nitrogen. The process is selected instead of carburizing, when high hardness with less distortion is required. But, because any steel that can be carburized can also be carbonitrided, the process introduces some of the benefits of nitriding (hard case, shallow case depths, lower processing temperatures, less distortion) to materials that cannot actually be nitrided.

Often carbonitriding is applied to low-cost, low-carbon steels where a wear-resistant surface is all that is required. Because the process is carried out at temperatures lower than those for conventional carburizing, carbonitriding is economical and allows more accurate distortion control. In addition, carbonitriding strengthens low-carbon steels subjected to light loads (where high-core properties are not required).

Vacuum Carburizing: This is a relatively new process where portions of the cycle are performed in a vacuum. The process, perhaps best described as partial-pressure carburizing, may produce a carburized case in less time than conventional carburizing. Of course, because of the vacuum requirement, equipment costs and processing costs may be higher than for conventional carburizing. Overall benefits of this process are yet to be established.

Carburizing can be followed by many other heat treatments. Carburizing followed by air-cooling, reaustenitizing, and then the conventional quench-and-temper can produce finer martensitic grains and eliminate much of the retained austenite. In addition, a number of carefully selected tempering cycles can further refine the structure. Naturally, each added cycle increases processing costs and must be justified by improved properties.

Parts may be selectively carburized. Surfaces can be "stopped off" by plating. Only those surfaces exposed to the carbonaceous atmosphere are hardened. A sequence often employed to selectively harden is (1) carburize the entire part, but do not harden; and (2) machine the carburized case off surfaces that are to remain soft. Only those surfaces where the case is left will be hard when reheated and quenched.

Nitriding: This is not new, but recent improvements in furnace design and control have moved

Chapter 2　Introduction of Mold Material

nitriding into the precision heat-treatment category. Nitriding is selected for precision parts where wear resistance and fatigue resistance are important.

Nitrogen, in the presence of certain catalysts and other elements, enters the surface of the metal and forms nitrides with some of the alloying metals. Few alloys can be nitrided: the "nitriding steels", some tool steels such as H−13, gray cast iron, and alloys such as the medium-carbon 4 100 series. In general, special alloys high in aluminum and chromium content are used for nitriding.

Nitriding occurs at reasonably low temperatures (about 900 °F, or 483 °C) and quenching is not required, so distortion is minimized. Usually, material is first heat-treated for desired core properties (often quench-and-tempered, for example, to a hardness that will yield satisfactory toughness but remain machinable). After finish machining, the part is nitrided. Finish grinding may follow nitriding, if necessary. Nitriding does not affect the previously established core properties. Sometimes, stress relieving before nitriding, followed by semifinish machining, can eliminate distortion completely.

Gas nitriding, by far the most precise process, requires preliminary surface preparation. Parts must be clean and either etched or grit-blasted for a specific surface texture. Improperly treated parts nitride unevenly and slowly.

 Words and Expressions

air-cool	v.	利用空气冷却；装置冷气设备
austenitize	v.	[冶]奥氏体化
carbonaceous	adj.	[化]碳的，碳质的，含碳的
compressive	adj.	有压缩力的
distortion	n.	扭曲；变形；曲解；失真
etch	v.	蚀刻
gradient	adj.	倾斜的
	n.	梯度，倾斜度，坡度
grit-blasted		喷钢砂处理的(有棱角的钢砂)
nitrid	n.	[化]氮化物
quench	vt.	结束；熄灭；淬火
	vi.	熄灭；平息
scuff	vi.	拖足而行；磨损
	vt.	以足擦地；践踏；使磨损
	n.	拖着脚走，磨损之处
vacuum	n.	真空；空间；真空吸尘器
	adj.	真空的；产生真空的；利用真空的
	vt.	用真空吸尘器打扫

课文导读 及 参考译文

（一）课文导读

本文从热处理过程和热处理理由这两方面介绍了钢的热处理。其中含铁材料的热处理包含几个重要操作：正火、球化、消除应力、退火、淬火、回火和表面硬化。

（二）课文参考译文

钢的热处理

热处理工艺技术要求是工程图纸中最重要的部分之一。适当的热处理是开发材料最佳潜在性能的有力工艺。通常，热处理被描述为对金属或合金按一定的速度进行加热和冷却的一种处理过程，以此来获得理想的性能。热处理主要用于使材料产生强化，但是某些热处理工艺使材料软化、韧化或提高其他性能。

金属或合金的内部由一种或多种原子所构成，这些原子在三维空间规则有序地堆砌形成晶体。晶体按照多种方式依次结合在一起，结合方式被称为显微结构或晶粒结构。由压缩、拉伸或剪切等作用力带来的塑性变形不但可以在某种程度上改变任何给定的物质结构，而且还能结合给定的热处理工艺获得较宽范围内的性能的变化。热处理工艺受时间、温度、温度变化速率和炉内空气的控制。具体热处理工艺的选择必须以成品零件所期望的性能为基础。

现今有大量用于不同目的的金属和合金。也有许多不同的热处理工艺。并非所有的热处理都能用于每种金属或合金。换句话说，选择的热处理工艺必须和指定材料所需要的热处理工艺相一致。不能不考虑材料而独立地选择热处理工艺。材料与热处理工艺一样重要。

热处理的原因

黑色金属。黑色金属工件热处理的原因有以下几点：消除内应力，通过细化晶粒度或使工件内部晶粒均匀来改变显微结构，通过添加或去除元素来改变表面化学状态，以及强化金属工件。

黑色金属的热处理。铁是工具钢中的主要成分，其中加入碳是为了使钢件硬化。加入合金元素是为了使钢件具有普通碳素钢所不具有的性能，例如在油或空气中淬硬的能力，增强的耐磨性，较高的韧性，和更大的硬化安全性。

含铁材料的热处理包含几个重要操作，常常被冠以如下这些称谓，例如正火、球化、消除应力、退火、淬火、回火和表面硬化。

正火是指将材料加热到临界温度范围以上 100 ℉ ~ 200 ℉（55 ℃ ~ 100 ℃）左右，然后在静止空气中冷却。这比通常的硬化温度高 100 ℉（55 ℃）左右。

正火的目的通常是为了细化锻造过程中被粗化的晶粒。对于大多数已合金化和未合金化的中碳锻造钢而言，强烈建议在锻造之后、机加工之前采用正火工艺以使材料的结构更加均匀，在大多数情况下这样可以提高材料的机械加工性。

高合金空气硬化钢从不需要正火，因为正火会使该类材料发生硬化而达不到初始目的。

球化处理是退火的一种形式，是在钢件的加热和冷却过程中生成圆形或球形的碳化物——钢中的硬质组成物。

工具钢通常进行球化处理以提高切削加工性。对于碳钢而言要加热到1 380 ℉~1 400 ℉（749 ℃~760 ℃），对于许多合金工具钢而言要加热到更高温度，然后恒温保持1~4小时，最后在炉内缓慢冷却。

消除应力。这是消除钢件在成型、冷加工、焊接或切削加工后冷却等过程中产生的内应力的一种方法。它是最简单的热处理工艺，只需要将材料加热到1 200 ℉~1 350 ℉（649 ℃~732 ℃），随后空冷或随炉冷却。

大型模具通常首先粗加工，然后消除应力和精加工。这样不但在切削加工过程中，而且在随后的热处理过程中都能使形状的变化减至最小。不同加热和冷却循环的组合以及横截面的改变，使得焊接区域也存在牢固的应力。这些应力在切削操作中会引起相当大的位移。

退火。退火过程常为将钢加热至高温并保持一定的时间，然后慢慢冷却。退火是为了达到均一化并且建立正常的平衡条件，以获取相应的性能。

购买的工具钢通常是退火态的。有时需要重新加工一个已经硬化的工具，那么这个工具必须经过退火处理。这种类型的退火处理是将钢加热到稍高于临界温度范围以上，然后非常缓慢地冷却。

淬火。该工艺是将钢加热到临界值以上，并以足够快的速度冷却通过临界温度范围，以使钢产生显著硬化。

回火。该工艺是将已淬火和已硬化的钢和合金加热到较低临界温度以下某一温度，以减少硬化时产生的内应力。

表面硬化。向钢质工件表面添加碳和随后的硬化操作是该类热处理的重要方面。这个过程包括使用熔融氰化钠混合物，用活性固体材料例如木炭、焦炭，或者用能够提供碳原子的气体、油类和干的氰化物填满渗碳介质。

（三）阅读材料（2）参考译文

表面硬化

表层硬化、局部硬化或表面硬化在铁质工件的心部周围生成坚硬的表层，表面硬化的工艺有几种，除感应硬化外，所有的表面硬化工艺都包括另一种材料扩散到金属表面中。

渗碳：最普通的扩散型硬化工艺是渗碳。简而言之，该工艺是指碳原子在高温（大约1 700 ℉或927 ℃）下扩散到金属中，实际上是将金属表面层从低碳钢转变为高碳钢。然后通过简单的淬火和回火使高碳表层硬化（因为硬度是碳含量的函数）。硬化表层耐磨损和划伤，而且因为存在有利的表面压应力而具有较高的疲劳抗力。表面拉应力对疲劳性能不利。

渗碳的技术要求必须包括对心部以及对表面的要求。精确控制的表面性能通常要求气体渗碳，因为可以精确监测含碳气氛。为了获得均匀的渗层深度和硬度，含碳气氛必须完全包围工件。当然，碳含量从渗层表面向心部逐渐降低，硬度的变化也是如此。有时将硬度测量值为洛氏硬度HRC50的位置定义为实际心部与表层或"有效表层"之间的分界。渗碳以后要尽可能少地从表面去除金属以防止去除掉硬化表层的最好部分；热处理后去除的渗碳层通常不超过20%。

渗碳工艺的典型技术要求包括：

1. 表面的表层硬度。

2. 有效表层深度。
3. 碳化物情况和限制。
4. 表层的硬度梯度。
5. 心部硬度。
6. 心部组织（全马氏体，无奥氏体，等等）。
7. 针对表面开裂的 NDT（检测）。
8. 渗碳后的最大切削量。

基于渗碳循环的一些变化被设计了出来，以满足某些特殊的需求：

碳氮共渗：碳氮共渗听起来像渗碳和渗氮的结合。实际上，碳氮共渗本质上是液态渗碳，但是它也包括氮的一些扩散。当要求高硬度并且小变形时，选择碳氮共渗而不选择渗碳。但是，因为能被渗碳的任何钢也能被碳氮共渗，所以该工艺给实际上不能被渗氮的材料带来了渗氮工艺的一些优点（表层硬、表层深度浅、加工温度低、变形小）。

碳氮共渗经常应用于只要求具备耐磨损表面的廉价低碳钢。因为碳氮共渗的温度低于传统渗碳的温度，所以碳氮共渗经济，而且可达到更精确的变形控制。此外，碳氮共渗可以强化承受轻载荷的低碳钢（不要求心部高性能的场合）。

真空渗碳：这是一种相对较新的工艺，其中部分循环在真空中进行。该工艺也许最好被称为部分压力渗碳。与传统渗碳相比，该工艺能在更短的时间内制备渗碳层。当然，由于真空要求，该工艺的设备成本和加工成本高于传统渗碳。该工艺全面的优点还有待建立。

渗碳后可以进行多种其他热处理。渗碳后空冷，再次奥氏体化，然后进行传统的淬火加回火，可以制备出细小的马氏体晶粒，并且消除大部分残余奥氏体。此外，许多经仔细选择的回火循环能进一步细化组织。每一个增加的循环必然会提高加工成本，而且必须通过提高的性能来证明是合算的。

工件可以有选择性地渗碳。可以通过电镀来遮盖表面，只有那些暴露在渗碳气氛中的表面被硬化。选择性硬化通常采用的顺序是：(1)对整个工件进行渗碳，但是不硬化；(2) 将需要保持软态的表面上的渗碳层切削掉。当再次加热和淬火时，只有那些保留了渗碳层的表面被硬化。

渗氮：渗氮不是一种新工艺，但是近来在加热炉设计和控制方面的进步使得渗氮成为精密热处理种类。对于重点要求耐磨性和疲劳抗力的精密工件选择渗氮工艺。

在有某些催化剂和其他元素的情况下，氮进入金属表面并与一些合金化元素形成氮化物。少数合金可以渗氮："氮化钢"、某些工具钢如 H-13、灰口铸铁和一些合金如中碳钢 4100 系列。含铝量和含铬量高的特殊合金通常用于渗氮。

渗氮在相当低的温度下进行（900 ℉或 483 ℃左右），不需要淬火，因此变形量最小。通常首先对材料进行热处理以获得期望的心部性能（例如通常进行淬火加回火，以使硬度达到能获得良好韧性但仍保持可切削加工性的程度）。切削加工完后对工件进行渗氮。如果需要，渗氮后可以精研磨。渗氮不会影响先前已获得的心部性能。有时渗氮前消除应力，随后半精加工能完全消除变形。

气体渗氮是最精确的工艺，要求进行表面预处理。工件必须清洁，对于特殊的表面结构必须进行蚀刻或者喷钢砂处理。处理不当的工件会渗氮不均匀而且渗氮缓慢。

Chapter 3

Press Process and Die Design
（冷冲压工艺及模具）

Lesson 5

Forming of Sheet Metals

Sheet metals are generally characterized by a high ratio of surface area to thickness. Forming of sheet metals is carried out generally by tensile forces in the plane of the sheet; otherwise the application of compressive forces could lead to buckling, folding, and wrinkling of the sheet.

In bulk deformation processes such as forging, rolling, extrusion, and wire drawing, there is an intentional change in the thickness or the lateral dimensions of the workpiece. However, in most sheet-forming processes any thickness change is due to the stretching of the sheet under tensile stresses (Poisson's ratio). Thickness decreases should generally be avoided as they could lead to necking and failure.

Method of Sheet Shearing

The shearing process involves cutting sheet metal by subjecting it to shear stresses, usually between a punch and a die much like a paper punch (Fig. 5-1). The punch and die may be any shape; they may be circular or straight blades similar to a pair of scissors. The major variables in the shearing process are: the punch force P, the speed of the punch, lubrication, surface condition and materials of the punch and die, their corner radii, and the clearance between the punch and the die.

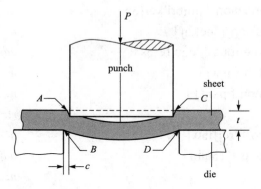

Fig. 5-1 Schematic illustration of the shearing process with a punch and die

In addition to the shearing processes described above, there are other techniques for cutting sheet metals. The sheet or plate may be cut with a saw, such as a band saw. This is a chip removal process. In another process, called shaving, the extra material from a rough sheared edge is trimmed by cutting (similar to removing a thin layer of wood with a chisel).

Flame cutting is another common method, particularly for thick steel plates. This process has wide applications in building ships and heavy structural components.

In friction sawing, a disk or blade rubs against the sheet or plate at surface speeds up to 25 000 ft/min (130 m/s) for disks, and 15 000 ft/min (80 m/s) for blades. In this process, the frictional energy is converted into heat, which then rapidly softens a narrow zone of the metal. The continuous movement of the disk or blade then pulls the softened metal out of the cutting zone. To help remove the material, some tools have teeth or notches. The friction sawing process is more suitable for ferrous alloys, as nonferrous metals have a tendency to adhere to the disk or blade and interfere with the cutting operation.

Words and Expressions

band sawn	n.	<美>[机]带锯
be due to		由于
buckle ['bʌkl]	v.	扣住；变弯曲
chip [tʃip]	n.	刮；修胡须；削，刨；刨花
chisel ['tʃizl]	n.	凿子，扁鏨
	v.	砍凿
flame [fleim]	n.	火焰；光辉，光芒
fold [fəuld]	vt.	折叠
forging ['fɔːdʒiŋ]	n.	锻炼；伪造
lubrication [ˌluːbri'keiʃən]	n.	润滑油
necking ['nekiŋ]	n.	颈缩；局部收缩
nonferrous ['nɔn'ferəs]	adj.	不含铁的，非铁的
notch [nɔtʃ]	n.	凹口，槽口
punch [pʌntʃ]	n.	冲压机；冲床；打孔机
	vt.	冲孔，打孔
ratio ['reiʃiəu]	n.	比；比率
rolling ['rəuliŋ]	adj.	旋转的，转动的
	n.	轧制；旋转
scissor ['sizə]	vt.	剪取；截取；删除；削减
	n.	(-s)剪刀
wrinkl ['riŋkl]	v.	(使)起皱纹

1. Sheet metals are generally characterized by a high ratio of surface area to thickness.

译文：金属板料的一般特征是表面积和厚度之间的比率很高。

解析：be characterized by：被赋予……的特征。a ratio of：……的比率。

2. Forming of sheet metals is carried out generally by tensile forces in the plane of the sheet; otherwise the application of compressive forces could lead to buckling, folding, and wrinkling of the

sheet.

译文：板料通常通过在板料面上施加拉力来成形，而如果使用压缩力就可能导致板料的弯曲、折叠和褶皱。

解析：此句为一个用连接词 otherwise 表转折的复合句，其中前句的主语是现在分词短语 forming of sheet metals。

3. However, in most sheet-forming processes any thickness change is due to the stretching of the sheet under tensile stresses (Poisson's ratio).

译文：然而，在大多数板料成形过程中，任何厚度的变化都是由于板料在拉应力作用下的延伸（泊松比）。

解析：be due to：由于，应归于。

4. The shearing process involves cutting sheet metal by subjecting it to shear stresses, usually between a punch and a die much like a paper punch (Fig. 5-1).

译文：冲裁过程是指在切应力的作用下对板料进行切割，通常是在一个凸模和一个凹模之间，极似纸张打孔器的操作（图5-1）。

解析：involve：包括，涉及，牵连。subject…to…：使某人或某物经历或遭受某事物。

5. In addition to the shearing processes described above, there are other techniques for cutting sheet metals.

译文：除了上述冲裁过程外，还有其他的板料切割方法。

解析：in addition to：加之……，除……之外。

6. In another process, called shaving, the extra material from a rough sheared edge is trimmed by cutting (similar to removing a thin layer of wood with a chisel).

译文：在另一个被称为缺口修整加工的过程中，通过切割对粗糙的冲裁边缘的多余材料进行切边（此过程就像用凿子凿去木材上薄薄的一层）。

解析：called shaving 前其实省略了 which is，合起来是作为修饰中心词 another process 的定语从句。

7. Flame cutting is another common method, particularly for thick steel plates. This process has wide applications in building ships and heavy structural components.

译文：火焰切割是另一种常见的方法，尤其适用于切割厚钢板。这种方法广泛地应用于造船和锻造巨型结构的零件。

解析：have application in：应用于某方面。

8. In friction sawing, a disk or blade rubs against the sheet or plate at surface speeds up to 25 000 ft/min (130 m/s) for disks, and 15 000 ft/min (80 m/s) for blades.

译文：在摩擦割锯中，锯盘或锯片摩擦表面板料或金属板的速度达到了锯盘为 25 000 英尺/分钟（130 米/秒），锯片为 15 000 英尺/分钟（80 米/秒）。

解析：rub：摩擦。rub 用于词组中表示"与……相摩擦"，其后一定要接介词 against。speed 在此句中为动词。句中 up to 意为"达到"。

9. In this process, the frictional energy is converted into heat, which then rapidly softens a narrow zone of the metal.

译文：在此过程中，摩擦能转换为热能，然后热能迅速软化金属板狭窄的部分。

解析：which 引导的是一个非限制性定语从句，修饰 heat。convert：转变，转换。soften：使变软，软化。

10. The friction sawing process is more suitable for ferrous alloys, as nonferrous metals have a tendency to adhere to the disk or blade and interfere with the cutting operation.

译文：摩擦锯割工艺更适合钢铁类合金，而有色金属在锯割时则有粘在锯片和刀片上的倾向，以至于妨碍切割过程。

解析：be suitable for：适合……。as 在句中的意思为"因为"，等同于 because。句中 have a tendency to 意为"有……的倾向"。

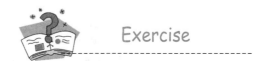

Fill in the blanks with the proper words.

1. However, in most sheet-forming processes any thickness change is _____ _____ the stretching of the sheet under tensile stresses.
2. Thickness decreases should generally be _____ as they could _____ _____ necking and failure.
3. The shearing process involves cutting sheet metal by _____ it _____ shear stresses, usually between a punch and a die much like a paper punch.
4. _____ _____ _____ the shearing processes described above, there are other techniques for cutting sheet metals.
5. The continuous movement of the disk or blade then _____ the softened metal _____ _____ the cutting zone.

（一）课文导读

金属板料的一般特征是表面积和厚度之间的比率很高，在加工过程中要避免厚度发生变化。板料的剪切加工方法有冲压机和冲模切割、带锯切割、火焰切割以及摩擦割锯。不同的切割方法有不同的适应性。

（二）课文参考译文

金属板料成形

金属板料的一般特征是表面积和厚度之间的比率很高。板料通常通过在板面上施加拉力来成形，而如果使用压缩力就可能导致板料的弯曲、折叠和褶皱。

在锻造、轧制、挤压和拉丝等体积变形过程中，工件的厚度或横向尺寸要求发生变化。然而，在大多数板料成形过程中，任何厚度的变化都是由于板料在拉应力作用下的延伸

（泊松比）。一般应该要避免厚度的减少，因为它可能导致颈缩与加工失败。

金属板料的冲裁方法

冲裁过程是指在切应力的作用下对板料进行切割，通常是在一个凸模和一个凹模之间，极似纸张打孔器的操作（插图5-1）。凸模和凹模可以是任何形状，它们可以是圆形的或类似于剪刀的笔直刀片。冲裁过程中的主要变量有：冲压力 P、冲压速度、润滑、表面状况、凸模和凹模的材料、圆角半径，以及凸模和凹模之间的间隙值。

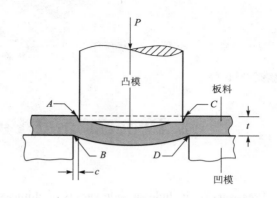

图 5-1　凸模与凹模冲裁工序示意图

除了上述冲裁过程外，还有其他的板料切割方法。可以用锯子，例如带锯，来切割板料或金属板。这是一种切屑过程。在另一个被称为缺口修整加工的过程中，通过切割对粗糙的冲裁边缘的多余材料进行切边（此过程就像用凿子凿去木材上薄薄的一层）。

火焰切割是另一种常见的方法，尤其适用于切割厚钢板。这种方法广泛地应用于造船和锻造巨型结构的零件。

在摩擦割锯中，锯盘或锯片摩擦表面板料或金属板的速度达到了锯盘为 25 000 英尺/分钟（130 米/秒），锯片为 15 000 英尺/分钟（80 米/秒）。在此过程中，摩擦能转换为热能，然后热能迅速软化金属板狭窄的部分。锯盘和锯片的持续运转使得这些软化的金属被拉到切割区之外。为了便于除去这些材料，一些工具做成齿和凹口。摩擦割锯工艺更适合钢铁类合金，而非铁金属在锯割时则有粘在锯片或刀片上的倾向，以至于妨碍切割过程。

Lesson 6

Press Process and Product Applications

Pressworking includes a wide variety of chipless processes by which workpieces are shaped from rolled metal sheets. A general term, stamping, is used almost interchangeably with pressworking.

Stampings are produced by the downward stroke of a ram in a machine called a press. The ram is equipped with special punches and moves toward and into a die block, which is attached to a rigid bed. The punches and the die block assembly are generally referred to as a "die set" or, more simply, as the "die". Pressworking operations are usually cold-worked at room temperature or in some cases hot-worked at an elevated temperature that does not exceed the annealing temperature of the metal.

Parts produced by pressworking operations can be as small as a shoe eyelet or as large as the end of a freight car. Compared to other metalworking processes, pressworking techniques offer an almost unlimited choice of metals and design versatility and can be produced in extremely large quantities. Metal stampings are lightweight, strong, and have a superior strength-to-weight ratio. It is estimated that the average household contains products in which there are over 100 000 pressworked items.

Skillful designers are often able to redesign parts previously made by forging or by casting, with significant savings in time, labor, and materials. Because it is practical to produce parts to close limits of accuracy, interchangeability is assured.

Pressworked parts are used for internal components on business machines, machine tools, household equipment, aircraft and small engines, and for locks, various other hardware items, and countless other functional applications. It is estimated that approximately half the weight of an automobile consists of pressworked parts: Formed parts are widely used as containers of various kinds, ranging from household pots and pans, to pails, buckets, and bins. Sheet-metal forms are also widely used for heating, exhaust, and ventilating equipment, medical and food processing equipment, for buildings and structures, household appliances (stoves, refrigerators, freezers, washers, and dryers), bathroom and plumbing and electrical articles, highway vehicles, farm equipment, office furniture, and for many other applications too numerous to mention.

Chapter 3 Press Process and Die Design

Words and Expressions

bathroom [ˈbɑːruːm]	n.	浴室，盥洗室	
bin [bin]	n.	箱柜	
bucket [ˈbʌkit]	n.	桶，一桶的量；铲斗	
countless [ˈkautlis]	adj.	无数的，数不尽的	
dryer [ˈdraiə]	n.	干衣机；干燥剂	
freight [freit]	n.	货物，船货；运费；货运	
	vt.	装货；运送	
furniture [ˈfəːnitʃə]	n.	家具；设备；储藏物	
pail [peil]	n.	桶，提桶	
pan [pæn]	n.	平底锅；盘子；面	
stamp [ˈstæmp]	v.	冲压	
unlimited [ʌnˈlimitid]	adj.	无限的；无约束的	
washer [ˈwɔʃə]	n.	洗衣人；垫圈；洗衣机；洗碗机	

1. Pressworking includes a wide variety of chipless processes by which workpieces are shaped from rolled metal sheets.

译文：冲压加工包括多种多样的无屑加工，轧制的金属板料通过无屑加工后成形为零件。

解析：a variety of = kinds of：多种多样的。shape：成形，做成某种形状；对……有重大影响。

2. The ram is equipped with special punches and moves toward and into a die block, which is attached to a rigid bed.

译文：滑块上装有专门的凸模向装在固定的工作台上的凹槽运动。

解析：be equipped with：配备有。which is attached to a rigid bed 是定语从句，修饰中心词 a die block。

3. The punches and the die block assembly are generally referred to as a "die set" or, more simply, as the "die".

译文：这些凸模和凹模装配组件一般被称为"模组"，或者更简单地被称为"冲模"。

解析：refer to：提及，谈到；与……有关；参考。

4. Pressworking operations are usually cold-worked at room temperature or in some cases hot-worked at an elevated temperature that does not exceed the annealing temperature of the metal.

译文：冲压加工的过程常常是在室温下进行的冷加工，或者某些情况下，是在不超过被加工金属的退火温度的较高温度下进行的热加工。

解析：cold-worked 为合成词，构词方式为形容词 cold+过去分词 worked，意为"冷加工"，其反义词为 hot-worked。

5. Parts produced by pressworking operations can be as small as a shoe eyelet or as large as the end of a freight car.

译文：冲压加工出来的零件可以如鞋带孔般小，也可以如运货汽车车厢般大。

解析：as small as：同……一样小。as... as... 意为"同……一样……"，表示同级比较，中间的形容词用原形。

6. Compared to other metalworking processes, pressworking techniques offer an almost unlimited choice of metals and design versatility and can be produced in extremely large quantities.

译文：与其他金属加工工艺相比，冲压技术在金属材料的种类及产品设计方面提供了非常广泛的选择空间，从而利用冲压技术可以生产大量的产品。

解析：compared to other metalworking processes 为过去分词短语作状语，意为"与其他金属加工工艺相比"。

7. Because it is practical to produce parts to close limits of accuracy, interchangeability is assured.

译文：由于制出与精确度十分接近的零件是可行的，所以零件间的互换性是可以保证的。

解析：此句中 it 为形式主语，真正的主语是后面的 to produce parts to close limits of accuracy。

8. It is estimated that approximately half the weight of an automobile consists of pressworked parts.

译文：据估计，汽车大约一半的重量都是由冲压零件组成的。

解析：it is estimated that：据估计。相同的说法还有：it is said that（据说）；it is reported that（据报道）；等等。

Exercise

Fill in the blanks with the proper words.

1. The punches and the die block assembly are generally _____ _____ as a "die set" or, more simply, _____ the "die".
2. Parts produced by pressworking operations can be _____ small _____ a shoe eyelet or _____ large _____ the end of a freight car.
3. Skillful designers are often _____ _____ redesign parts previously made by forging or by casting, with significant savings _____ time, labor, and materials.
4. Pressworked parts are used _____ internal components _____ business machines, machine tools, household equipment, aircraft and small engines, and _____ locks, various other hardware items, and countless other functional applications.
5. Formed parts are widely used _____ containers of various kinds, ranging _____ household

pots and pans, _____ pails, buckets, and bins.

（一）课文导读

冲压加工是指利用压力机滑块的下行冲程制作出冲压件的过程。冲压技术可以生产出大量大小不同的产品。一般的家用产品中就有许多是通过冲压加工制成的。冲压加工成形的产品有广泛的应用性。取暖、排气和通风设备，医疗和食品加工仪器，建筑机器，家用器具，管道设施，电器，交通工具，农用机器等都有冲压加工产品的应用。

（二）参考译文

冲压工艺和产品应用

冲压加工包括多种多样的无屑加工，轧制的金属板料通过无屑加工后成形为零件。"冲压"是一个概括性的术语，几乎可以与冲压加工替换使用。

压力机滑块的下行冲程制作出冲压品。滑块上装有专门的凸模向装在固定的工作台上的凹槽运动。这些凸模和凹模装配组一般被称为"模组"，或者更简单地被称为"冲模"。冲压加工的过程常常是在室温下进行的冷加工，或者某些情况下，是在不超过被加工金属的退火温度的较高温度下进行的热加工。

冲压加工出来的零件可以如鞋带孔般小，也可以如运货汽车车厢般大。与其他金属加工工艺相比，冲压技术在金属材料的种类及产品设计方面提供了非常广泛的选择空间，从而利用冲压技术可以生产大量的产品。金属冲压品重量轻、坚固，并且具备较高比强度。据估计，一般家庭所拥有的产品中，有超过十万的产品是通过冲压加工做成的。

技术熟练的设计者们通常能重新设计先前锻造或铸造生产的零件，大大地节省了时间、劳力和原材料。因为制出与精确度十分接近的零件是可行的，所以零件间的互换性是可以保证的。

冲压加工可以制作出商用机器、机器工具、家用设备、飞行器和小型发动机的内部零件，也可以为锁和其他各种各样的五金器具以及无数实用的功能型器械制作零件。据估计，一辆汽车大约一半的重量都是由冲压零件组成的。冲压后成形的部件广泛地用作各种各样的容器，如家用的壶与锅、提桶、水桶和箱柜。金属板料成形件也广泛地被用在取暖、排气和通风设备，医疗和食品加工机械，建筑机械，家用器具（火炉、冰箱、冰柜、洗衣机和干衣机），浴室和管道设施，电器，交通工具，农用机械，办公设备以及其他应用设施，因为太多，在此不能——列举。

Lesson 7

Classification of Dies

Dies may be conveniently classified according to their function: those used to cut metal and those used to form metal. Cutting operations include blanking, trimming, shaving, cutoff, shearing, piercing, slitting, perforating, lancing, extruding, notching, and nibbling. Forming operations include a variety of processes, which are usually grouped under the general headings of bending or forming or squeezing or drawing.

Conventional Dies

These consist of one or more mating pairs of rigid punches and die blocks. Additional auxiliary equipment may be added to increase the pressworking versatility. Conventional dies include single-operation or simple dies, compound or combination dies, progressive dies, transfer dies, and multiple dies.

Some die sets are designed to perform a single pressworking operation, which may include any of the operations listed under cutting or forming. Such dies are called single-operation dies or simple dies. One operation is accomplished by a single stroke of the press. Fig. 7-1 illustrates a simple die for trimming a horizontal flange on a drawn shell. In this example, the workpiece is positioned on a locating plug. After the ring-shaped scrap from a sufficient number of trimmed shells has accumulated, a scrap ring at the bottom is severed with each stroke of the press by the scrap cutter and falls clear.

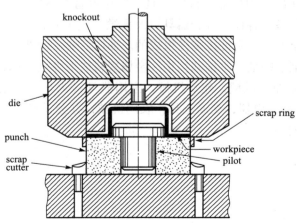

Fig. 7-1 The tooling for trimming a horizontal flange on a drawn shell

Chapter 3 · Press Process and Die Design

Compound or Combination Dies

These will be used when two or more operations are performed at one station. Fig. 7-2 illustrates a part that was formed by one stroke of the press in a compound die. In this example, the dished washer was simultaneously blanked, pierced, and formed. In mass-production operations compound dies are more economical than a series of single operation dies, and they are usually more accurate.

Progressive Dies

These are used for high-production applications. In this case coil stock or flat strip is fed from station to station. The die performs work at some or all of the stations during each press stroke. When all the work has been completed, the work is cut off and unloaded. Fig. 7-3 shows an adapter ring that was produced in a five-station progressive die. Progressive dies are generally expensive to construct. In addition, the cost of the required auxiliary equipment is high. Progressive dies are usually set up on automatic presses with a scrap cutter, feeder, straightener, and uncoiler.

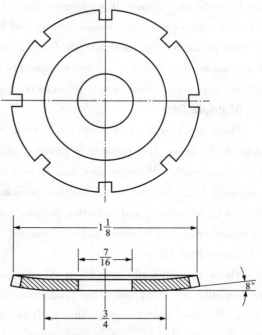

Fig. 7-2 An example of a part formed in a compound die by one stroke of the press

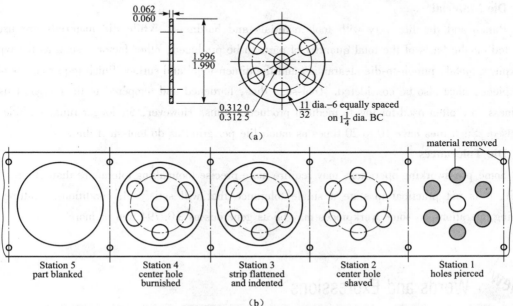

Fig. 7-3 An adapter ring formed in a five-station progressive die
(a) adaptor ring blank; (b) 2% wide strip

Transfer Dies

These are used to produce parts which, because of their general configuration, are difficult to feed from station to station in progressive dies. Individual precut blanks are first prepared by feeding coil stock into a press. Additional cutting and forming operations on the separate workpiece blanks are then performed by mechanically feeding each blank from station to station. Like progressive dies, transfer dies and their related equipments (presses, special attachments, and feeding devices, etc) are expensive. Their use is recommended only in cases of high-quantity production.

Multiple Dies

These are also used in mass production. Such dies produce two or more workpieces at each stroke of the press. It is possible to produce pairs of right-hand and left-hand parts, duplicate parts, or unrelated parts. Multiple dies may consist of two or more single-operation dies or multiples of compound dies. Advantages of multiple dies may include savings in material resulting from more efficient blank layout, and reduction of labor costs. The leading disadvantages are increased costs in die construction and in setup and maintenance.

Short-Run Dies

These are frequently used by metal fabricators, particularly for blanking operations. Such dies are generally employed when the production run is limited from a few hundred to about 10 000 pieces. Because short-run dies can usually be made more quickly and installed in the press with less setup time than that is required for conventional dies, they are used for trial runs and as a means of expediting delivery of parts. In most cases, changes can be readily incorporated into the design of conventional high-production die sets. Inexpensive steel-rule and template dies are the principal types used for short-run applications.

Die Material

Punch and die life vary with tool materials and hardness. While die materials are usually selected on the basis of the total quantity of parts to be produced, other factors, such as the type of workpiece metal, punch-to-die clearance, and the dimensional and surface finish requirements of the workpiece, must also be considered. Tool-steel dies, hardened and tempered to their highest usable hardness, are often used for low to medium production runs. However, for longer runs, carbide dies are used. Such dies have 10 to 20 times as much life per grind as do tool-steel dies.

Die Tolerances

Some pressworking operations may require more precise toolmaking tolerances than others. As a result, it is not practical to state a single tolerance that will satisfy all conditions. Cutting and forming operations on some workpieces may be as generous as ±0.794 mm or more.

Words and Expressions

accumulate [əˈkjuːmjuleit]   v.   积聚；堆积

blank ['blæŋk]	v.	打孔；盖印
compound ['kɔmpaund]	n.	混合物；[化]化合物
	adj.	复合的
	v.	混合；配合
duplicate ['djuːplikeit]	adj.	复制的；副的
	n.	复制品，副本
	vt.	复写，复制；使加倍
horizontal [ˌhɔri'zɔntl]	adj.	地平线的；水平的
lance [lɑːns]	n.	标枪；长矛
	v.	切开
nibble ['nibl]	n.	细咬，轻咬，啃
	v.	分段冲裁
perforate ['pəːfəreit]	v.	打孔
pierce [piəs]	vt.	刺穿，刺破，穿透
progressive [prə'gresiv]	n.	进步论者
	adj.	前进的；(税收)累进的
recommend [rekə'mend]	vt.	推荐，介绍使受欢迎；托付
slit [slit]	vt.	切开；撕裂
	n.	裂缝；狭长切口
trim ['trim]	n.	修整
versatility [ˌvɜːsə'tiləti]	n.	多功能性

Notes

1. Dies may be conveniently classified according to their functions: those used to cut metal and those used to form metal. Cutting operations include blanking, trimming, shaving, cutoff, shearing, piercing, slitting, perforating, lancing, extruding, notching, and nibbling.

译文：模具可以根据它们的功能很方便地进行分类：一些用于使金属分离，一些用于使金属成形，分离工序包括落料、修剪、切边、切断、冲裁、冲小孔、割缝加工、冲孔、切缝、模压和开槽。

解析：be classified into：被分为……。与之意思相同的短语还有 be grouped into，be placed into。

2. Some die sets are designed to perform a single pressworking operation, which may include any of the operations listed under cutting or forming.

译文：一些冲压模被设计用来执行单工序冲压操作，可以是分离或成形工序的任一操作。

解析：be designed to：被设计成……。which may include any of the operations listed under cutting or forming 是修饰中心词 a single pressworking operation 的定语从句。

3. In this example, the dished washer was simultaneously blanked, pierced, and formed.

译文：在这个例子中，中凹垫圈的落料、冲小孔、成形工序都是同时完成的。

解析：simultaneously：（adv.）同时地，同时发生地。

4. Compound dies are more economical in mass-production operations than for a series of single operation dies, and they are usually more accurate.

译文：在大规模生产过程中，复合模比一系列单工序模具更为经济，而且通常还精确得多。

解析：more...than：比起……更加……。

5. Progressive dies are usually set up on automatic presses with a scrap cutter, feeder, straightener, and uncoiler.

译文：通常是在自动压力机上装上碎边剪切废料装置、进料器、矫直机和开卷机来装配成连续模。

解析：set up：组装，装配。

6. In most cases, changes can be readily incorporated into the design of conventional high-production die sets.

译文：大多数情况下，可将试生产中所发生的变化很快地用来作为设计传统高产量冲压模的依据。

解析：be incorporated into：被混合成……，被具体化为……。

7. Punch and die life vary with tool materials and hardness.

译文：凸模与凹模的使用寿命因工具的材料和硬度的不同而不同。

解析：vary with：因……而改变，随……而变化。

8. While die materials are usually selected on the basis of the total quantity of parts to be produced, other factors, such as the type of workpiece metal, punch-to-die clearance, and the dimensional and surface finish requirements of the workpiece, must also be considered.

译文：模具材料的选择主要是依据生产产品的数量来决定的，其他的因素，如产品的类型、凸模和凹模之间的间隙、产品的尺寸以及表面的粗糙度要求也必须考虑。

解析：on the basis of：以……为基础。sth is considered = take sth into consideration：考虑。

9. Tool-steel dies, hardened and tempered to their highest usable hardness, are often used for low to medium production runs.

译文：小批量生产和中批量生产通常采用加硬和调节至最高可用硬度的工具钢制作模具。

解析：hardened and tempered to their highest usable hardness 是过去分词短语作定语，修饰中心词 tool-steel dies。

10. As a result, it is not practical to state a single tolerance that will satisfy all conditions.

译文：因此，确定一个可以满足所有条件的单一公差是不可行的。

解析：本句中的 it 为形式主语，真正的主语是后面的不定式 to state a single tolerance that will satisfy all conditions。

Chapter 3 Press Process and Die Design

Fill in the blanks with the proper words.
1. Conventional dies _____ _____ one or more mating pairs of rigid punches and die blocks.
2. In this example, the workpiece is positioned _____ a locating plug.
3. In this case coil stock or flat strip is fed _____ station _____ station.
4. The leading disadvantages are increased costs _____ die construction and _____ setup and maintenance.
5. However, _____ _____ _____, carbide dies are used.

（一）课文导读

冲压模具可以根据它们的功能分为分离工序模具和成型工序模具。常规模具由一副或多副成对的固定冲床和模块组成，包括单工序或简单模具、复合或组合模具、连续模具、移动模具和一模多件模具。本文主要介绍了各类模具的构成和操作，以及模具材料和模具公差方面的情况。

（二）课文参考译文

冲压模的分类

模具可以根据它们的功能很方便地进行分类：一些用于使金属分离，一些用于使金属成形。分离工序包括落料、修剪、切边、切断、冲裁、冲小孔、割缝加工、冲孔、切缝、模压、开槽。成形工序包括多种工艺过程，这些工艺过程通常包括弯曲、成形、挤压和拉伸。

常规模具

常规模具由一对或多对成对匹配的刚性凸模与凹模组成。也可以增加额外的辅助设备来扩充冲压加工的功能。常规模具包括单工序或简单模具、复合或组合模具、连续模具、移动模具和一模多件模具。

一些冲压模被设计用来执行单工序冲压操作，可以是分离或成形工序的任一操作。这些冲模叫做单工序模具或简单模具，即压力机的一次冲程完成一个工序。图7-1是一个拉深的壳体进行水平凸缘切边的简单冲模。在这个例子中，

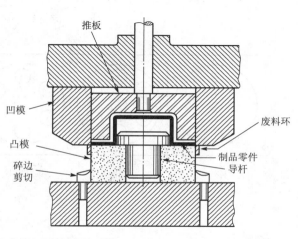

图7-1 在拉伸外壳上进行水平凸缘修剪的模具

加工件被安放在一个定位销上。当壳体零件被切边后的环形废料堆积到一定的量后,废料切刀在压力机的每次冲程下切断底部的环形废料并掉落。

复合或组合模

当在一个工位上要完成两个或多个操作时,就要运用到复合模或组合模具。图 7-2 所示为复合模在压力机的一次冲程过程中冲压成形一个零件的情形。在这个例子中,中凹垫圈的落料、冲小孔、成形工序都是同时完成的。在大规模生产过程中,复合模比一系列单工序模具更为经济,而且通常还精确得多。

连续模(级进模)

连续模应用于大批量生产。在这种情况下,带卷或板料从一个工位传递到下一个工位。在压力机的每一次冲程过程中,模具都会在某些或所有的工位上完成操作。当所有的操作完成之后,工件被切断而分离。图 7-3 所示为由 5 个工位构成的连续模制作一个连接环零件的情形。连续模的制造费用一般很高。另外,必要的辅助设备的成本也很高。通常是在自动压力机

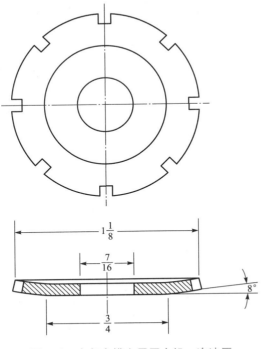

图 7-2　在复合模上用压力机一次冲压成形零件的实例

上装上碎边剪切废料装置、进料器、矫直机和开卷机来装配成连续模。

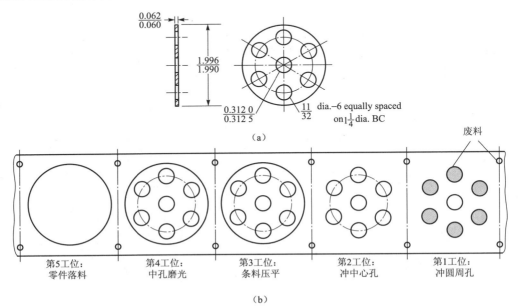

图 7-3　在 5 个工位连续模上成形的连接环
(a) 接合环坯件;(b) 2% 宽幅的条料

移动模

移动模具用来生产那些由于自身结构的原因而很难在连续模中由一个工位传递到下一个工位的零件。将带卷送入压力机先制作单个的落料坯件,再用机械将各个分离的坯件一个工位一个工位地送进,完成余下的剪切和成形工序。与连续模具类似,移动模具及其相关设备(压力机、专门的附加装置、进料装置等)也是昂贵的,所以只有在大规模生产的情况下,才推荐使用移动模具。

一模多件模具

一模多件模具也适用于大批量生产。这种模具在压力机的一次冲程过程中就可以制作出两个或多个零件。一模多件模具可以制作成对的左右对称零件、重复零件或者不相干的零件。它可以由两个或多个单工序模具,或者由若干复合模具组成。一模多件模具的优点是能节约材料,提高材料利用率和减少劳力成本。它主要的缺点就是增加了模具制造以及调整和维修的成本。

小批量生产模具

金属件制作者经常使用此类模具,尤其是用它来进行落料操作。当生产产量限制在几百到一万件左右时,一般采用小批量生产模具。因为与常规模具相比,制造一个小批量生产模具更快,将其安装到压力机上所需的时间也比常规模具更短。它们可以用来进行试生产,也可以作为一种加快零件制作过程的方法。大多数情况下,可将试生产中所发生的变化很快地用来作为设计传统高产量冲压模的依据。便宜的钢材和模具模板是小批量生产模具应用的主要类型。

模具材料

凸模与凹模的使用寿命因工具的材料和硬度的不同而不同。模具材料的选择主要是依据生产产品的数量来决定的,其他因素,如产品的类型、凸模和凹模之间的间隙、产品的尺寸以及表面的粗糙度要求也必须考虑。小批量生产和中批量生产通常采用加硬和调节至最高可用硬度的工具钢制作模具。但是也可采用碳化物质模具,其使用时间更长。这类模具的使用周期是工具钢模具的10~20倍。

模具公差

与其他加工工艺相比,冲压加工的操作对于工具制造也许要求更加精确的公差。因此,确定一个可以满足所有条件的单一公差是不可行的。一些制件的分离和成形工序也许要求其公差在±0.794毫米或更大范围之内。

Lesson 8

Presses

The main function of a stamping press is to provide sufficient power to operate the die and the movements necessary to close and open the die. The press also maintains the alignment of the punches and dies. Some simple single-action presses can be operated by hand feeding the metal stock and by manually tripping the dies on a one-to-one basis. Most production stamping presses, however, are equipped with multiple station dies and are designed for high-speed, completely automatic stock feeding and stamping operations. Some presses are restricted to single-purpose types of operations, such as bending, coining, or punching. Multipurpose presses are available and, depending upon the design and construction of the dies, are capable of performing varied stamping operations, such as cutting, forming, and punching.

Press Classification

Presses are classified in various ways. Some manufacturers list presses according to the types of work for which the press has been designed. Other manufacturers group press types according to the methods of power transmission, which may include simple manual operation or mechanical, steam, or hydraulic power operation. Frame and bed designs constitute another way that may be used to classify stamping presses. Finally, press types may be listed according to the action or the numbers of rams (single, double, triple); the methods of ram operations (crank, eccentric, toggle, knuckle-level); and the positions of the ram guides (vertical, horizontal, inclined).

JIC Press Identification System

The Joint Industry Conference (JIC) system of identifying press characteristics is in general use. In a typical sample, S4-750-96-72, the press is identified by the S as a single-action model (D is used for double-action, T for triple-action, and OBI for open-back inclinable); by the 4 as having four-point suspension; by the 750 as being rated at 750-ton (670-metric-ton) capacity; and by the 96 and 72 as having a bed measuring 96 in (2.45 m) left to right and 72 in (1.83 m) front to back. Any other press can be so identified by substitution of appropriate numerals for the number of suspension points, tonnage rating, and bed dimensions.

The JIC also recommends that a metal tag be attached permanently to the press, stating the stroke length, shut (or maximum die opening) height, kind and length of adjustment, strokes per minute, and bed size and machine weight.

Chapter 3 Press Process and Die Design

Inclined Press

An inclinable open-back press with a gap frame is shown in Fig. 8-1. This press, though shown in the vertical position, can be tilted backwards to permit the parts and scrap to slide off the back side. Parts can slide by gravity into a tote box, or material may be fed by chute into the dies. Most presses of this type are adjustable and vary from vertical to a steep-angle position. Inclinable presses are often used in the production of small parts involving bending, punching, blanking, and similar operations.

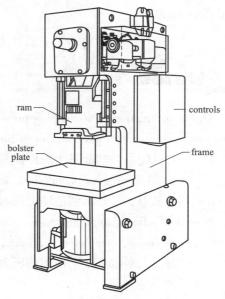

Fig. 8-1 An inclinable open-back press

 ## Words and Expressions

alignment [ə'lainmənt]	n.	队列；结盟
chute [ʃuːt]	n.	瀑布；斜道
crank [kræŋk]	n.	[机]曲柄；奇想
	vt.	[机]装曲柄
	vi.	[机]转动曲柄
eccentric [ik'sentrik]	adj.	古怪的
	n.	行为古怪的人
inclined [in'klaind]	adj.	倾向……的
knuckle ['nʌkl]	n.	关节；铰链接合
manually ['mænjuəli]	adv.	用手
multiple station dies		多工位模
substitution [ˌsʌbsti'tjuːʃən]	n.	代替；取代作用；代入法；置换
toggle ['tɔgl]	n.	[海]绳针；套索钉
	vt.	拴牢
tonnage ['tʌnidʒ]	n.	登记吨位；排水量
triple ['tripl]	n.	三倍数；三个一组
	adj.	三倍的
	vt.	三倍于
	vi.	增至三倍

vertical [ˈvəːtikəl]　　　　　　　　　　adj. 垂直的；直立的；顶点的
　　　　　　　　　　　　　　　　　　　n. 垂直线；垂直面；竖向

Notes

1. The main function of a stamping press is to provide sufficient power to operate the die and the movements necessary to close and open the die.

译文：压力机的主要功能是为操作模具和开、闭模具所需的运动提供足够的动力。

解析：by hand feeding the metal stock and by manually tripping the dies on a one-to-one basis 为两个并列的方式状语，修饰谓语动词 operate。

2. Most production stamping presses, however, are equipped with multiple station dies and are designed for high-speed, completely automatic stock feeding and stamping operations.

译文：然而，大多数生产上使用的压力机常常被安装上多工位的模具，用于高速的、完全自动化送料的冲压过程。

解析：此句的主体部分为 presses are equipped with multiple station dies and are designed for stock feeding and stamping operations，此句中有两个并列的谓语。

3. Some presses are restricted to single-purpose types of operations, such as bending, coining, or punching.

译文：有些压力机局限于进行单一功能的操作，如弯曲、压印或冲孔。

解析：restrict：限制，限定。

4. Multipurpose presses are available and, depending on the design and construction of the dies, are capable of performing varied stamping operation, such as cutting, forming, and punching.

译文：多工序压力机的功能取决于模具的设计和制造，它们能够完成各种冲压工序，例如落料、成形以及冲孔。

解析：be capable of：能够。

5. Other manufacturers group press types according to the methods of power transmission, which may include simple manual operation or mechanical, steam, or hydraulic power operation.

译文：其他制造商则根据驱动方式的不同对压力机进行分类，包括简单人力驱动或机械驱动、气动或液压驱动。

解析：句中 group 为动词，意为"分类"。which may include simple manual operation or mechanical, steam, or hydraulic power operation 是定语从句，修饰中心词 the methods of power transmission。

6. The JIC also recommends that a metal tag be attached permanently to the press, stating the stroke length, shut (or maximum die opening) height, kind and length of adjustment, strokes per minute, and bed size and machine weight.

译文：英国联合工业委员会还推荐在永久地贴于压力机的标签上标出以下内容：压力机的行程、闭模（或最大开模）高度、可调整的类型和长度、每分钟冲压的次数、床身尺寸以及冲床的重量。

解析：句中 a metal tag be attached permanently to the press 为动词 recommend 的宾语从句，

宾语从句中的谓语应该为"should+动词原形"或直接用动词原形。此外，句中的 stating the stroke length, shut (or maximum die opening) height, kind and length of adjustment, strokes per minute, and bed size and machine weight 在宾语从句中是现在分词短语作状语，用于表示和从句中的动词 be attached to 同时发生的动作。

7. This press, though shown in the vertical position, can be tilted backwards to permit the parts and scrap to slide off the back side.

译文：尽管这种压力机显示的是垂直位置，但它仍可以向后倾斜，以便零件和碎料沿着后部滑落。

解析：此句可以改写成：Though this press is shown in the vertical position, it can be tilted backwards to permit the parts and scrap to slide off the back side.

8. Most presses of this type are adjustable and vary from vertical to a steep-angle position.

译文：大多数这种类型的压力机是可以调整的，而且可以把它们从垂直位置调至倾角位置。

解析：adjustable：可调整的。

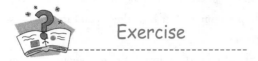

Fill in the blanks with the proper words.

1. Some simple single-action presses can be operated _____ hand _____ the metal stock and _____ manually _____ the dies on a one-to-one basis.
2. Some presses are _____ _____ single-purpose types of operations, such as bending, coining, or punching.
3. Any other press can be so _____ by substitution _____ appropriate numerals _____ the number of suspension points, tonnage rating, and bed dimensions.
4. Parts can _____ by gravity _____ a tote box, or material may be _____ by chute _____ the dies.
5. Most presses of this type are _____ and vary _____ vertical _____ a steep-angle position.

Reading Materials (3)

Drive Mechanisms for Presses

Most of the drive mechanisms for transmitting power to the slide are shown in Fig. 8-2. The most common drive is the single crank, which gives a slide movement approaching simple harmonic motion. On a downstroke the slide is accelerated. Reaching its maximum velocity at midstroke, it is then decelerated. Most press operations occur near the middle of the stroke at maximum slide velocity. The eccentric drive gives a motion like that of a crank and is often used where a shorter stroke is required. Its proponents claim it has greater rigidity and less tendency for deflection than a crank drive. Cams are used where some special movement is desired such as a swell at the bottom of the stroke. This drive has some similarity to the eccentric drive except that roll followers are used to transmit the motion to the slide.

Rack and gear presses are used for applications requiring a very long stroke. The movement of the slide is slower than in crank presses, and uniform motion is attained. These presses have stops to control the stroke length and may be equipped with a quick-return feature to raise the slide to starting position. The arbor press is a familiar example.

Hydraulic drive is used in presses for a variety of work. It is especially adapted to large pressures requiring slow speed for forming, pressing, and drawing operations.

In the screw drive, the slide is accelerated by a friction disk that engaged the flywheel. As the flywheel moves down, greater speed is applied. From beginning to end of the stroke the slide motion is accelerated. At the bottom of the stroke the amount of stored energy is absorbed by the work. The action resembles that of a drop hammer, but it is slower and there is less impact. Presses of this type are known as percussion presses.

Several link mechanisms are used in press drives because of the motion or because of their mechanical advantage. The knuckle joint mechanism (Fig. 8-2) is common. It has a high mechanical advantage near the bottom of the stroke when the two links approach a straight line. Because of the high load capacity of this mechanism, it is used for coining and sizing operations.

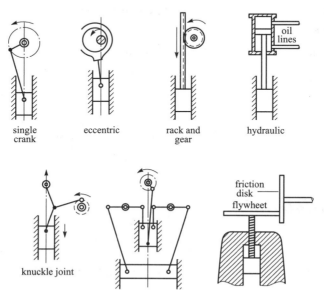

Fig. 8-2　Drive mechanisms used on presses

Eccentric or hydraulic drives may be substituted for the crank shown in the figure. Toggle mechanisms used primarily to hold the blank on a drawing operation are made in a variety of designs. The auxiliary slide in the figure is actuated by a crank, but eccentrics or cams may be used. The principal aim of this mechanism is to obtain a motion having a suitable dwell so that the blank can be held effectively.

An application for the microprocessor in control of a press is in stretch forming. About 75% of aircraft frames are stretch formed. A microprocessor that has memory capabilities can be programmed by using an acceptable part that has been formed manually. Under a playback mode any number of parts can be produced. Another advantage of the microprocessor is that programs can be memorized and transferred to tapes or disks that serve as permanent records.

Words and Expressions

cam	n.	凸轮
gear	n.	齿轮；传动装置
	v.	调整；(使)适合；换挡
harmonic	adj.	和谐的；和声的；融洽的
	n.	谐波；和声；谐函数
microprocessor	n.	[计]微处理器
percussion	n.	敲打；打击乐器
playback	n.	录音重放；录音重放装置；重放
rack	n.	行李架
velocity	n.	速度，速率；周转率

课文导读及参考译文

（一）课文导读

压力机的主要功能是为操作模具和开、闭模具所需的运动提供足够的动力。它的分类方法很多，根据压力机的工作类型、驱动方式、结构与床身设计等等，都可以对压力机进行分类。英国联合工业委员会压力机的特征识别方法是一种普遍常用的方法，该识别方法主张用恰当的数字表示悬架点数目、吨位级别和床身尺寸来表征任何压力机。

（二）课文参考译文

压力机

压力机的主要功能是为操作模具和开、闭模具所需的运动提供足够的动力。压力机也保持了凸模与凹模的相对位置的正确。一些简易的单动压力机可以用手动操作进料，也可以手动将模具在两个相对的模座之间移动。然而，大多数生产上使用的压力机常常被安装上多工位的模具，用于高速的、完全自动化送料的冲压过程。有些压力机局限于进行单一功能的操作，如弯曲、压印或冲孔。多工序压力机的功能取决于模具的设计和制造，它们能够完成各种冲压工序，例如落料、成形以及冲孔。

压力机的分类

压力机有多种分类方法。有些制造商根据压力机设计时实现的工作的类型将其进行分类。其他制造商则根据驱动方式的不同对压力机进行分类，包括简单的人力驱动或机械驱动、气动或液压驱动。按结构与床身设计分类是压力机的另一种分类方法。总之，压力机可以按照其动作或滑块的数目（单动、双动、三动）；滑块工作方法（曲柄、偏心曲柄、肘杆式、铰接式）以及滑块导轨的位置（垂直、水平、倾斜）进行分类。

英国联合工业委员会的压力机识别方法

英国联合工业委员会压力机的特征识别方法是一种普遍常用的方法。S4-750-96-72 是一个典型的例子，此类压力机被鉴定为单动模式的压力机，代号为 S（D 代表双动，T 代表三动，OBI 代表开式向后压力机）；用 4 表示有四点式悬架；750 是指公称压力为 750 英吨（670 公吨）；96 和 72 分别表示床身左右尺寸 96 英寸（2.45 米）前后尺寸 72 英寸（1.83 米）。可以用恰当的数字表示悬架点数目、吨位级别和床身尺寸来表示其他任何压力机。

英国联合工业委员会还推荐在永久地贴于压力机的标签上标出以下内容：压力机的行程、闭模（或最大开模）高度、可调整的类型和长度、每分钟冲压的次数、床身尺寸以及冲床的重量。

可倾式压力机

图 8-1 表示的是一个装有间隙框架的开式向后可倾式压力机。尽管这种压力机显示的是垂直位置，但它仍可以向后倾斜，以便零件和碎料沿着后部滑落。零件能在重力的作用下滑入一个零件箱，或者也运用斜道向模具进料。大多数这种类型的压力机是可以调整的，而且可以把它们从垂直位置调至倾角位置。可倾式压力机常用于小型零件的生产，如弯曲、冲孔、落料和类似的其他工序。

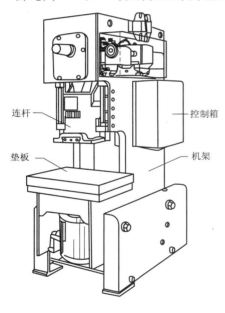

图 8-1 开式向后可倾式压力机

（三）阅读材料（3）参考译文

压力机的驱动机构

图 8-2 表示的是大多数用来给滑块传送动力的驱动机构。单曲柄是最普通的驱动装置，它给滑块提供能使其达到谐波运动的动力。下冲程使滑块加速。中冲程使滑块达到最大速度，之后滑块开始减速。大多数冲压操作是在压力机处于中冲程、滑块速度最大的时候进行的。偏心曲柄驱动装置的传动方式与单曲柄驱动的方式相似，常常在要求使用较短冲程的时候使用此种机构。偏心曲柄驱动机构的支持者们认为与单曲柄驱动机构相比，此种机构的刚

性更大，偏差的可能性更小。当要求执行一种诸如冲程结束时强度增强的特别的运转动作时，就要使用凸轮。这种驱动方式与偏心曲柄驱动类似，除了它要使用辊式的从动部件向滑块传送动力以外。

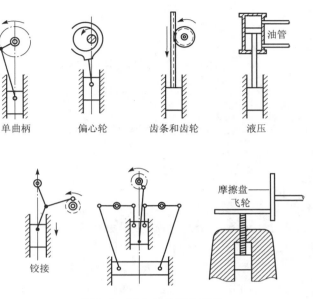

图 8-2　压力机的驱动装置

齿条和齿轮压力机用来执行要求很长冲程的操作工序。这种压力机的滑块运转速度慢于曲柄压力机，并且滑块达到的是匀速运动。这类压力机上设有挡块以控制行程长度，并装备有将滑块升至启动位置的急回装置。杠杆式冲床就是一个常见的例子。

液压驱动可以运用于各种用途的压力机。它尤其适用于要求慢速成形、压平和拉深操作的大压力冲压。在螺旋驱动机构中，连接飞轮的摩擦盘使滑块加速。当飞轮往下运转时，滑块速度就更快。从冲程的开始到结束，滑块的运转不断加速。在冲程结束之时，储存的能量就能被吸收运用到加工当中。这种运动方式与落锤类似，但是它的速度更慢，冲击力更小。这种类型的压力机被称之为打击压力机。

一些连杆机构由于其运动方式或机械优势而被运用于压力机驱动机构中。曲柄连杆机构（图 8-2）的应用十分普遍。当两个连杆的位置即将成一条直线时，冲程末端就有很大的机械优势。由于这种机构具有很大的负载量，它通常用于压印和整形加工。

如图所示，偏心曲柄驱动机构或液压驱动机构可以替代曲柄驱动机构。肘杆机构主要用来在拉深工序中给法兰压边，其设计多种多样。在图中，辅助滑块虽然是由一个曲柄驱动，但是也可能用偏心轴或凸轮驱动。这种机构的主要目的是为了得到一个适当的保压动作，以使法兰能得到有效的压边。

控制压力机的微处理器可运用于拉深成形工序。约有 75% 的飞机机架都是拉深成形的。可以运用已经人工制作出来的合格的零件对拥有记忆能力的微处理器进行编程。压力机在重放模式下就可以生产出任何数量的零件。微处理器的另一个优点在于它能记住程序并且可以使其转换为磁带或碟片作为永久的记录。

Lesson 9

Shear Operation

Cutting metal involves stressing it in shear above its ultimate strength between adjacent sharp edges as shown in Figure 9-1.

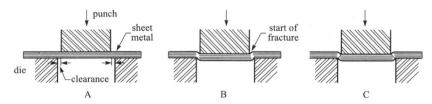

Fig 9-1   Process of shearing metal with punch and die

A. Punch contacting metal; B. Plastic deformation; C. Fracture complete

As the punch descends upon the metal the pressure first causes a plastic deformation to take place as in Fig. 9-1B. The metal is highly stressed adjacent to punch-and-die edges, and fractures start on both sides of the sheet as the deformation continues. When the ultimate strength of the material is reached, the fracture progresses, and, if the clearance is correct and both edges are of equal sharpness; the fractures meet at the center of the sheet as shown in Fig. 9-1C. The amount of clearance, which plays an important part in die design, depends on the hardness of the material. For steel it should be 5% to 8% of the stock thickness per side. If improper clearance is used, the fractures will not meet, and instead must cross the entire sheet thickness, using more power.

Flat punches and dies as shown in the figure require a maximum of power. To reduce the shear force the punch or die face should be made at an angle, so that the cutting action is progressive. This distributes the shearing action over a greater length of the stroke and can reduce the power required by up to 50%.

Blanking, as shown in Fig. 9-2, is the operation of cutting out flat areas to a desired shape. It is usually the first step in a series of operations. In this case the punch should be flat and the die given some shear angle so that the finished part will be flat. Punching or piercing holes in metal, notching metal from edges, or perforating are all similar operations. For these operations the shear angle is on the punch and the metal removed is scrap. Trimming is the removal of "flash" or excess metal from around the edges of a part and is essentially the same as blanking. Shaving is similar except that it is a finishing or sizing process.

Chapter 3 Press Process and Die Design

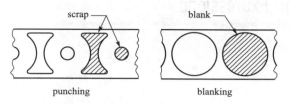

Fig. 9-2 Illustrating the difference between punching and blanking operations

Slitting is making incomplete cuts in a sheet as illustrated in Figure 9-3. If a hole is partially punched and one side bent down as a louver, it is called lancing. All these operations may be done on presses of the same type and differ little except in the dies that are used.

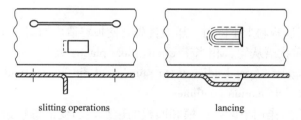

Fig. 9-3 Examples of slitting and lancing operations

The formula for the pressure P (in pounds-force) that is required to blank or punch a material, assuming there is no shear angle on the punch or die, is expressed as

$$P = SL_t$$

and for round holes,

$$P = \pi DS_t$$

where

S — Shear strength of materials, psi (Pa)

L — Sheared length, in (mm)

D — Diameter, in (mm)

t — thickness of materials, in (mm)

Because of the severe service encountered, die parts are made from carbide. When the punch becomes dull, usually after about 150 000 punches, it is reground. The small high-speed presses used with these dies employ coiled stock and operate at 300 spm. After the press operation the punched material is recoiled and hardened. Then follows grinding, honing, and stropping. Finally, the blades are snapped apart, inspected, and, depending upon brand, are coated with Teflon and packaged.

Words and Expressions

clearance ['kliərəns]	n.	公差；清除
formula ['fɔːmjulə]	n.	公式；规则；客套语
fracture ['fræktʃəriŋ]	n.	破裂，碎裂；龟裂；(水力)压裂
grinding ['graindiŋ]	n.	磨；摩擦；碾
honing ['həuniŋ]	n.	珩磨，搪磨

Notes ▶▶

1. As the punch descends upon the metal the pressure first causes a plastic deformation to take place as in Fig. 9–1B.

译文：当凸模在金属板料上下降时，压力首先会使板料造成如图9–1B所示的塑性变形。

解析：as 引导时间状语从句，相当于 when。take place：发生。

2. The metal is highly stressed adjacent to punch-and-die edges, and fractures start on both sides of the sheet as the deformation continues.

译文：金属在高应力作用下紧贴凸模和凹模的边缘，当变形继续进行时，板料与凸、凹模接触的两个边缘都将产生裂纹。

解析：此句为一个并列句，用连接词 and 连接。as the deformation continues 中的 as 意为"当……的时候"。

3. When the ultimate strength of the material is reached, the fracture progresses, and, if the clearance is correct and both edges are of equal sharpness; the fractures meet at the center of the sheet as shown in Figure 9–1C.

译文：当板料达到了它的极限抗拉强度时，裂纹就会产生，而如果公差正确，且两边的边缘同样锋利时，裂纹就会出现在如图9–1C所示中的板料中央。

解析：此句是由连接词 and 连接的两个并列句，前句是一个包含由 when 引导的时间状语从句的复合句，后句是一个由 if 引导的条件状语从句的复合句。

4. The amount of clearance, which plays an important part in die design, depends on the hardness of the material.

译文：在模具设计中起着重要作用的间隙值取决于原材料的硬度。

解析：此句中 which plays an important part in die design 是定语从句，修饰中心词 the amount of clearance。

5. To reduce the shear force the punch or die face should be made at an angle, so that the cutting action is progressive. This distributes the shearing action over a greater length of the stroke and can reduce the power required by up to 50%.

译文：为了减小冲裁力，凸、凹模之间应该被加工出一定的角度，以便于使切削成为一个渐进的过程。这有利于将冲裁过程分布在更长的凸模下降行程中，从而可以将冲裁力减小 50%。

解析：to reduce the shear force 为目的状语。distribute ... over ...：把……分布于……。

6. Punching or piercing holes in metal, notching metal from edges, or perforating are all similar operations.

译文：在金属中冲孔或冲小孔、从边缘精密冲裁金属或打孔都属于类似的操作。

解析：此句为简单句，句子的主语部分由并列动名词 punching、piercing holes in metal、notching metal from edges 和 perforating 构成。

7. Because of the severe service encountered, die parts are made from carbide.

译文：模具零件的材质是碳化物，因为它具有高度的耐用性。

解析：encounter：相遇（尤指意外遇到）；遭遇。

8. Finally, the blades are snapped apart, inspected, and, depending upon brand, are coated with Teflon and packaged.

译文：最后，刃口被折断，接受检查，并且根据其类型在其表面涂上特氟纶然后进行包装。

解析：be coated with：被涂上，覆盖上。

Fill in the blanks with the proper words.

1. As the punch descends _____ the metal the pressure first causes a plastic deformation _____ take place _____ in Fig. 9-1B.
2. Blanking, as shown in Fig. 9-2, is the operation of cutting _____ flat areas _____ a desired shape.
3. Trimming is the _____ of "flash" or excess metal from around the edges of a part and is _____ the same as blanking.
4. All these operations may be done _____ presses _____ the same type and differ little _____ _____ the dies that are used.
5. _____ _____ the severe service encountered, die parts are made _____ carbide.

Reading Materials (4)

Structure of Stamping Die

The question of setting the punch means that this detail is clamped in the ram and then the latter is lowered first to allow the punch to slide into the die opening. Next the press setter must gradually move the die around using feelers to ensure that the clearance between the punch and the aperture is the same all round. The die is then clamped to the machine bed. This is not an easy task and one that becomes increasingly more difficult as the blank profile becomes more irregular in shape. However, if the material being blanked is thick then the larger die and punch clearance required does give some degree of latitude, but if the clearance is made too great in an endeavour to assist in press setting, then the amount of burr thrown up is likely to be objectionable and would possibly require removing at some later stage. It should be noted here that clearances and blanking pressures are dealt with in a later chapter.

The Follow-on Tool

The follow-on or progression-type tool is undoubtedly the most popular class of blanking tool and the piercing of several holes can be incorporated in the process. However in the example shown in Fig. 9-4 an opportunity is taken to include exactly the same item as used for the open-style of tool, but this time the tool incorporates what is known as a trigger stop as a means of positioning the strip as it passes through the guides. The details are fitted to a top and bottom bolster having the usual pillars and bushes for location purposes and although there is some resemblance to the previous tool, the design is obviously more expensive to construct and is only used when the quantity justifies the cost.

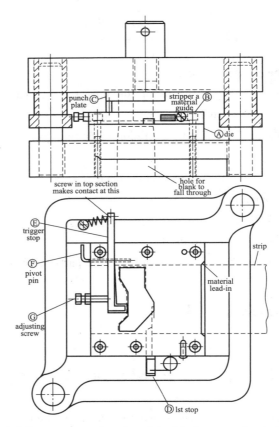

Fig. 9-4　A follow-on type of tool used in this instance simply for blanking

With the open type of press tool the wastage of a small piece of material when the strip is initially inserted in the guides is not of great importance and the blanking punch on descending will clip away a tiny section before the strip passes on to the stop. However, in the case of a follow-on tool the first stop is included to just allow the punch to clip only enough to ensure that a clean face reaches the second stop, and usually this first stop is spring-loaded as shown in the illustration. The material is now fed along to the second or trigger stop that always remains down until the top section descends, and an adjustable screw in that part strikes the protruding end and causes the stop to swing upward to allow the stock to pass underneath when feeding forward. Further explanation of this and other stops are given in a later chapter.

In both these tools the severed blank drops through the bottom generally into a container, but when many thousands are blanked a stacking arrangement is incorporated in the press cycle and the parts are gathered together in convenient stacks for easy removal from the presses.

Fig. 9-4 also indicates how a die-set is chosen according to the size of a die and in this case a cast-iron lower bolster will suffice as the thickness is sufficient for this type of component. Slots to locate both the die and punch plate are shaped in the bolsters. The details are ground on the side edges to ensure they are a push fit in the grooves, and a single dowel in each case provides the lateral location, and the author contends that these slots are a valuable asset to a toolmaker especially during the final fitting of the details on their respective bolsters because a positive location is secured by the grooves and only the dowelling is needed to complete the final assembly. Components blanked in this type of tool require only a press having about 25 mm stroke so on this occasion there is no fear that the pillars will lose contact with the bushes.

The scrap material is disposed of in three ways. For short strips that are not too difficult to handle, these are merely collected and sent to the scrap bin, but cropping them to short pieces can be effected by scrap choppers, and these are ideal if the material that emerges from the press is thin, tangly and likely to cause injury to an operator. The third method is to wind the scrap into a roll in a scrap winder.

Blanking a component which is large or made from a material not available in coil or strip, requires a different type of tool because with the pieces already to a shape — square, rectangular or perhaps some odd form — these will need locating on the die before the punch descends, and this means that the material will require holding flat prior to the punch contacting the surface.

Fig. 9-5 shows a simplified version of such a construction and although the blank is circular, the method is, of course, applicable to all other shapes. The die A is located in a recess bored in the lower bolster, but the pressure ring-cum-stripper B is attached to the upper section and surrounds the punch C, but is prevented from falling away against the spring pressure by three or four long screws. The punch is also recessed into the bolster in exactly the same way as the die is positioned.

A series of small locating pins D are fitted to the die and because the accuracy in locating the pre-cut blank is immaterial to within possibly ±0.5 mm, filing small flats on these pins is quicker than attempting a machining operation. The material is thus dropped between the pins, the ram set in motion and when the pressure ring contacts the material this is pushed down flat just before the punch descends a little further and compresses the springs to pass into the die and cut a blank. On

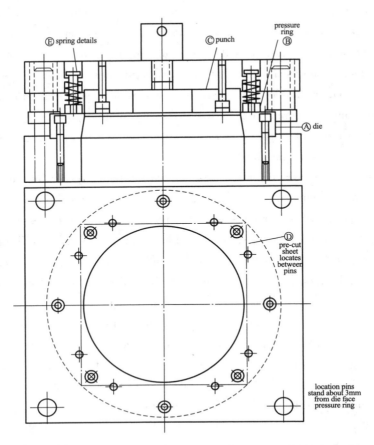

Fig. 9-5 Blanking large components requires a tool of equally massive proportions, especially if the blanked detail must fall through the lower bolster. In this example the disc is circular and this means a large hole in the base member. The material is positioned between pins and a spring loaded pad holds in flat on the die during the actual blanking stage.

ascending the ring holds the material against the die face until the punch has withdrawn into the ring and so it leaves the material on the die and prevents it clinging to the punch where it would prove an awkward operation to remove. The cut blank in the meantime has already fallen through the die in the normal manner, but care is essential with large blanks to see that the hole in the machine base is sufficiently large for this purpose. One or two points of interest concern whether the die is large enough in diameter to overcome any tendency to burst and the bolster is made from a thick piece of steel.

Words and Expressions

ascend	v.	攀登，上升
bolster	n.	垫子

	v.	支持
clamp	n.	夹子,夹具,夹钳
	vt.	夹住,夹紧
dispose	vi.	处理事务;作决定
	vt.	布置,安排
endeavour	n.	<英>尽力,竭力
feeler	n.	试探器;探针;测隙规
groove	n.	(唱片等的)凹槽
	vt.	开槽于
objectionable	adj.	引起反对的,讨厌的
odd	adj.	奇数的,单数的;单只的,不成对的;临时的,不固定的;带零头的,余的
recess	n.	(墙壁等的)凹进处;[解]隐窝
	vt.	使凹进
	vi.	休假,休息
stacking	n.	堆垛;分层
stripper	n.	清除的人;剥离器
trigger	vt.	引发,引起,触发
	n.	扳柄;制动器

(一)课文导读

金属板料冲裁在其极限强度之上对板料施加应力并使板料在两个锋利的相邻刀口之间进行冲压。落料工序是指在板料上裁剪出想要的形状。冲裁工艺对间隙值、凸模、冲压力等等有很严格的要求。本文举例说明了冲裁过程,并提供了计算冲压力的相关公式。

(二)课文参考译文

冲裁工序

如图 9-1 所示,金属板料的冲裁通常是指在其极限强度之上对板料施加应力并使板料在两个锋利的相邻刃口之间进行冲压的过程。

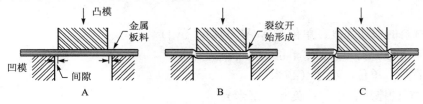

图 9-1 运用凸模与凹模冲裁板料的过程

当凸模在金属板料上下降时,压力首先会使板料造成如图9-1B所示的塑性变形。金属在高应力作用下紧贴凸模和凹模的边缘,当变形继续进行时,板料与凸、凹模接触的两个边缘都将产生裂纹。当板料达到了它的极限抗拉强度时,裂纹就会产生,而如果公差正确,且两边的边缘同样锋利时,裂纹就会出现在如图9-1C中所示的板料中央。在模具设计中起着重要作用的间隙值取决于原材料的硬度。对于钢板而言,其单面间隙值应该是其材料厚度的5%~8%。如果间隙值不当,裂纹就不会重合,且必须要用更大的力量来使裂纹穿透整个板厚。

图中所示的凸模和凹模要求具备最大的冲裁力。为了减小冲裁力,凸、凹模之间应该加工出一定的角度,以便使切削成为一个渐进的过程。这有利于将冲裁过程分布在更长的凸模下降行程中,从而可以将冲裁力减小50%。

图9-2所示的落料工序是指在板料上裁剪出想要的形状。它通常是一系列工序的第一步。在这种情况下,凸模应当是平的,而凹模应做成斜刃才能使制作出的零件是平整的。在金属中冲孔或冲小孔、从边缘精密冲裁金属或打孔都属于类似的工序。在这些工序中,斜刃设在凸模上,切除的金属是废料。修整工序是指切除零件边缘上闪光或多余的金属,基本上与落料工序类似。类似的还有刮削工序,不过它是一个精冲或园整过程。

切口工序如图9-3所示是指在板料上进行不完整的切割。如果一个孔只有部分被冲切,而板料的一边向下弯曲形成天窗状,就叫做切舌。所有这些工序都可以使用同类型的压力机,只是使用的模具稍微不同。

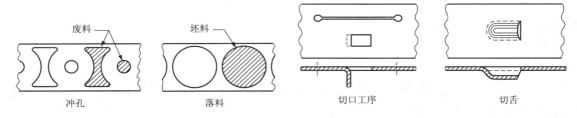

图9-2 冲孔与落料工序的区别　　　　图9-3 切口与切舌

假设凸模或凹模上没有斜刃角,对材料进行落料或冲孔所需压力P(以磅为单位)的公式如下:

$$P = SL_t$$

冲圆孔所需压力的公式为

$$P = \pi DS_t$$

式中,

S——材料的剪切强度,单位:磅/平方英尺(帕)

L——冲裁轮廓线的长度,单位:英寸(毫米)

D——直径,单位:英寸(毫米)

t——材料的厚度,单位:英寸(毫米)

模具零件的材质是碳化物,因为它具有高度的耐用性。凸模在运行了150 000次之后通常就会变钝,此时就要重新研磨。使用这些模具的小型高速压力机采用带卷,每分钟冲程次数为300。冲压材料在冲压之后收缩变硬。然后再进行磨削、搪磨和磨剃操作。最后，刃口被折断，接受检查，并且根据其类型在其表面涂上特氟纶然后进行包装。

（三）阅读材料（4）参考译文

冲裁模的结构

放置凸模的关键在于用连杆滑块夹紧凸模，然后连杆先下降使凸模滑入凹模槽内，随后，凸模安装员必须逐渐移动凹模，运用塞规确保凹模与凸模之间四周的间隙都是相同的。接着凹模被夹紧在床身上。这不是一项容易的任务，而且坯料外形越不规则，这项任务就会变得越困难。然而如果要落料的材料很厚，那么所需的凹模与凸模间隙将有一定程度的放大范围。但如果为了尽量适应压力机的安装而形成过大的间隙，那么坯料上就会产生大量的毛刺，并且还有可能需要在接下来的工序中清除它们。间隙以及冲裁将在下面的章节进行阐述。

连续模

连续模或级进模毫无疑问是最普遍的冲裁模具，它能合并完成多孔冲孔操作。在图9-4所示的例子中，连续模恰好也包含了开式模所使用的相同的零件。但在此例中，模具与被称为自动挡料销的零件结合在一起，成为在条料通过导料板时定位条料的工具。这种零件被安装在顶部和底部的模垫上，模垫上装备有常见的用来定位的导柱和导套，虽然此类模具与之前的模具有一些类似，但很明显此类设计的制造成本更高，而且也只有在生产数量所带来的效益能保证成本的前提下才会使用此种模具。

在开式模具中，当条料开始送入导料板时，小片材料的损耗并不重要，并且落料凸模在下压时将在条料传递到挡料销之前对条料进行小部分的修剪。但在连续模中，首先会有一个始用挡料销使凸模恰好充分修剪条料，以确保条料到达第二个挡料销时表面清洁。如图所示，始用挡料销通常用弹簧加载。材料开始顺延进入到第二个自动挡料销时，挡料销一直保持向下，直到上模下降，并且这个部件上的调整螺钉能打击突出的一端，促使挡料销向上旋转，使板料在进料时能在其底下通过。后面的章节将对此以及其他挡料销进行进一步的说明。

两种模具中，切断的落料件一般都是通过底部孔落入一个容器，但是当要求冲压成千上万的工件时，就要安排压力机的周转循环装置，当零件堆积到一定程度时便从压力机上移开。

图9-4也说明了怎样根据凹模的规格来选择模座，在这个例子中，只需要一个铸铁的下模垫就足够了，因为它的厚度能充分满足此类元件。用来定位凹模与凸模板的止扣在模垫上成形。这些零件的各个面的边缘被碾磨以确保它们作紧配合，实际上，单单一个销子就能进行侧面定位，作者认为对于模具制造者来说，尤其是当他们在各自的模垫上对零件进行最后的装配时，这些定位止扣是非常重要的。因为止扣能保证有效的定位，而且也只需要销来完成最后的装配工作。在这类模具上落料元件只要求压力机具备25毫米的冲程，所以在这种情况下，也就不必担心导柱与导套相脱离。

处理废料的方法有三种。短的易于处理的条料只需对其进行收集并送入废料箱，也可以用碎边剪切机将废料切成小块。如果来自压力机的废料薄且缠绕在一起，并且有可能对操作人员造成伤害时，这两种方法都是很理想的。第三种方法是将废料放入废料卷筒拧成筒状。

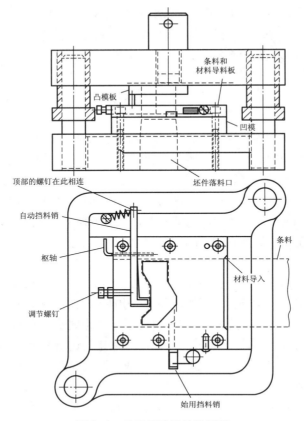

图 9-4 只用于冲孔的连续模

　　落料大元件或者材料不能卷曲或剥离的元件时，就要用到一种不同类型的模具，因为随着零件已经成形——方形、矩形或其他奇怪的形状——在凸模下压之前需要将它们定位在凹模上，这也就是要求这种材料在凸模冲压其表面之前保持平整的原因。

　　图 9-5 是此类模具结构的一个简易形式。虽然图中的坯料是圆形的，但这种方法对于所有其他形状的坯料都是适用的。凹模 A 安置在下模垫的止扣处，但是弹性卸料板 B 被安装在较高的部位，并将凸模 C 包围，同时装设有 3 或 4 个长螺钉防止其因弹性压力而脱落。此外凸模也按照与凹模同样的安装方式安装在模垫的止扣处。

　　凹模上安装了一系列的定位销 D，因为定位事先切割坯料的精确度的公差在±0.5毫米之内才不会造成实质的影响，在这些定位销上定位小型平板料要快于机器操作。因此材料跌落在定位销之间，连杆运转，并且当卸料板接触材料时，材料在凸模正要进一步下压并压缩弹簧之前被平整下推至凹模，从而冲裁出一个坯件。凸模回升时，卸料板保持材料面对凹模面直至凸模退回至卸料板之内，这样就使材料留在凹模上，并防止其粘住凸模，否则就要对其执行难以操作的卸料工序。同时切割的坯料正常地通过凹模落料。冲压大的坯料一定要注意机床上的洞是否大得足以进行出料。还有一两个问题值得关注，即凹模的直径是否大得足以防止任何破裂的可能，以及模垫是否要采用厚的钢质材料。

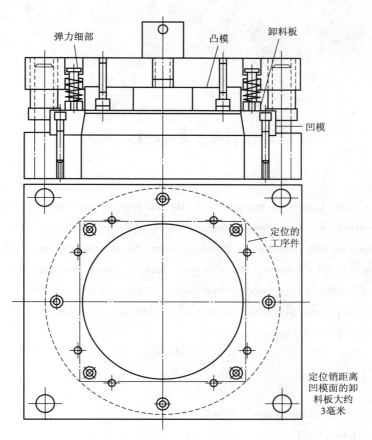

图 9-5 采用垫板脱模的大型落料模具示意图。此例子用的是圆形盘料，意味着底部存在一个大口。在落料操作进行时，材料被放在模具的定位销和弹性卸料板之间保持平整

Lesson 10

Bending Operation

Bending and forming may be performed on the same equipment as that used for shearing — namely, crank, eccentric, and cam-operated presses. Where bending is involved the metal is stressed in both tension and compression at values below the ultimate strength of the material without appreciable changes in its thickness. As in a press brake, simple bending implies a straight bend across the sheet of metal. Other bending operations, such as curling, seaming, and folding, are similar, although the process is slightly more involved. Bending pressures can be determined using the following empirical relationship:

$$F = \frac{1.3LSt^2}{W}$$

where

F—Bending force required, tons (MPa)

L—Length of bend, in (mm)

S—Ultimate tensile strength, tons/in^2 (Pa/mm^2)

t—Metal thickness, in (mm)

W—Width of V-channel, or U lower die, in (mm)

A V-channel or a U-die is a punch-and-die arrangement. These punch and dies often form a 90° internal angle between the faces of the metal. The empirical constant of 1.33 is a die-opening factor proportional to metal thickness.

In designing a rectangular section for bending, one must determine how much metal should be allowed for the bend, because the outer fibers are elongated and the inner ones shortened. During the operation the neutral axis of the section is moved toward the compression side, which results in more of the fibers being in tension. The thickness is slightly decreased, and the width is increased on the compression side and narrowed on the other. Although correct lengths for bends can be determined by empirical formulas, they are influenced considerably by the physical properties of the metal. Metal that has been bent retains some of its original elasticity, and there is some elastic recovery after the punch is removed as shown in Fig. 10-1. This is known as springback.

Fig. 10-2 shows a generalized stress-strain curve that explains the importance of the yield point in bending operations. If a part A-B is stretched to a length of A-C and released, it will return to its original length A-B. If A-B is stretched past the proportional limit L on the stress-strain curve to a length A-D and the load removed, then its final length is A-E. The springback in this case is E-D. The line M-E is parallel to L-B. Thus, depending on the stress-strain characteristics of the metal and the load employed, an indication of the extent of springback is possible.

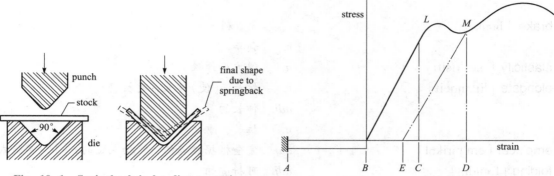

Fig. 10-1 Springback in bending operations

Fig. 10-2 Springback and its relationship to stress-strain

Springback may be corrected by overbending an amount such that when the pressure is released, the part will return to its correct shape. Springback is more pronounced in large-bend radii. The minimum-bend radius varies according to the ductility and thickness of the metal.

A forming die, designed to bend a flat strip of steel to a U-shape, is shown in Figure 10-3. As the punch descends and forms the piece, the knockout plate is pressed down, compressing the spring at the bottom of the die. When the punch moves up, the plate forces the work out of the die with the aid of the spring. Such an arrangement is necessary in most forming operations because the metal presses against the walls of the die, making removal difficult. Parts that tend to adhere to the punch are removed by a knockout pin that is engaged on the upstroke.

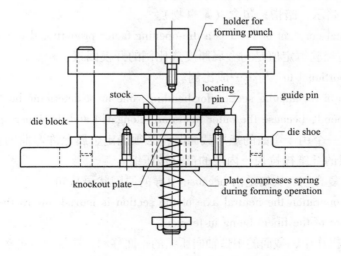

Fig. 10-3 Forming punch and die

Words and Expressions

brake [breik]	n.	闸,刹车
	v.	刹车
elasticity [iˈlæsˈtisiti]	n.	弹力;弹性
elongate [ˈiːlɔŋgeit]	v.	拉长,(使)伸长,延长
	adj.	伸长的
	n.	拉长,伸长
empirical [emˈpirikəl]	adj.	完全根据经验的,经验主义的;[化]实验式
folding [ˈfəuldiŋ]	adj.	可折叠的
neutral [ˈnjuːtrəl]	n.	中立者;齿轮的空挡
	adj.	中立的;中性的
pronounced [prəˈnaunst]	adj.	讲出来的;显著的;断然的;明确的
retain [riˈtein]	vt.	保持;保留
seaming [ˈsiːmiŋ]	n.	缝合,接合,缝拢,合拢
springback	n.	回弹
yield point	n.	屈服点

 Notes

1. As in a press brake, simple bending implies a straight bend across the sheet of metal.

译文:在压弯机中,简单的弯曲操作是指对金属板料的直角弯曲操作。

解析:imply:暗示,暗指;包含(某事物)。

2. The empirical constant of 1.33 is a die-opening factor proportional to metal thickness.

译文:1.33 的经验恒定值是与金属厚度成比例的开模因素。

解析:be proportional to:与……成比例。

3. In designing a rectangular section for bending, one must determine how much metal should be allowed for the bend, because the outer fibers are elongated and the inner ones shortened.

译文:在设计一个用于弯曲的矩形截面时,必须确定金属允许弯曲的金属的量,因为板料在弯曲时,外层的纤维被拉长,内层的部分被压缩。

解析:此句为复合句,从句是由 because 引导的原因状语从句。

4. During the operation the neutral axis of the section is moved toward the compression side, which results in more of the fibers being in tension.

译文:在这项操作中,截面的中性轴向压缩的一面移动,这就使得更多的纤维得到拉伸。

解析:此句中 which 引导的是定语从句。result in:结果是,导致。

5. Although correct lengths for bends can be determined by empirical formulas, they are influenced considerably by the physical properties of the metal.

译文：尽管弯曲的正确长度可由经验公式决定，但它还深受金属的物理特性的影响。

解析：此句为复合句，从句是由 although 引导的让步状语从句。be influenced by：受……的影响。

6. Thus, depending on the stress-strain characteristics of the metal and the load employed, an indication of the extent of springback is possible.

译文：因此，根据金属的压力应变特征和所受的载荷，就有可能确定回弹的范围。

解析：句中 depending on the stress-strain characteristics of the metal and the load employed 是现在分词短语作状语。

7. Springback may be corrected by overbending an amount such that when the pressure is released, the part will return to its correct shape.

译文：可以通过适当地加大弯曲量来纠正回弹，以保证压力释放后，零件能够恢复到其正确的形状。

解析：be corrected by：通过……纠正。such that：以至于。

8. A forming die, designed to bend a flat strip of steel to a U-shape, is shown in Fig. 10-3.

译文：图 10-3 所示是用来将平钢带弯曲成 U 型件的成形模具。

解析：designed to bend a flat strip of steel to a U-shape 是作后置定语，修饰 a forming die。

9. Parts that tend to adhere to the punch are removed by a knockout pin that is engaged on the upstroke.

译文：用于上冲程中的顶杆可以卸下容易粘住凸模的零件。

解析：engage：聘用，雇佣；参加或从事（某事）。

Fill in the blanks with the proper words.

1. Bending and forming may be performed _____ the same equipment _____ that used _____ shearing — namely, crank, eccentric, and cam-operated presses.
2. During the operation the neutral axis _____ the section is moved _____ the compression side, which results _____ more of the fibers being _____ tension.
3. Metal that has been bent _____ some of its original _____, and there is some elastic _____ after the punch is removed as shown in Fig. 15.1.
4. When the punch moves up, the plate forces the work _____ _____ the die _____ the aid _____ the spring.
5. Parts that tend to _____ _____ the punch are removed by a knockout pin that is engaged _____ the upstroke.

Reading Materials (5)

An Example of Grouping, Piercing and Bending

An example with the workpiece reversed in the tool is shown in Fig. 10-4; in addition, holes are pierced while the blank is also produced so giving an increase in output. The tool suffers from the same problem, however, that a slight springback can occur and is not curable. However providing this point is not of great significance the design is a good one. The strip is passed through in the manner shown in the "blanking tool" chapter, to push against the stop pin E and the descending punch, which has the dual function of initially cutting the strip to length and then performing the bending operation, crops the piece which is held between the face of punch N and pad G. Further movement carries the blank down over the edge of the die D and so bending up the end as shown in the drawing.

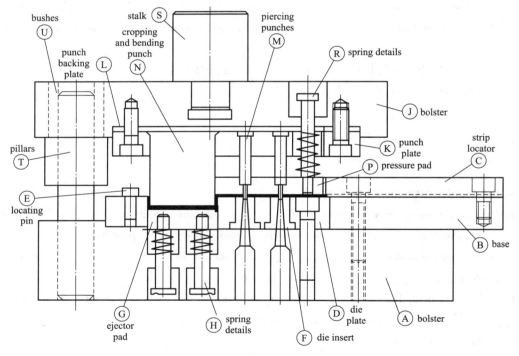

Fig. 10-4 Grouping, piercing and bending are accomplished on this tool and to secure full benefit from this form of production, an incliable press fitted with an air blast for quick removal of a detail is desirable

The first blank is wasted, unless some form of stop is added to position the strip and so allow both punches to pierce the necessary holes in the initial blank, but the holes appear in any subsequent cropped off piece so the bent component emerges with the holes correctly placed. The strip incidentally is held flat against the die surface by the pressure pad P after it has been pushed through the strip locator C.

A tool of this type, where the finished component is deposited on the die surface when pressing is completed, is best produced on an inclinable press using an air blast to ensure that the part is clear before the press ram again descends.

Words and Expressions

deposit	n.	堆积物；沉淀物；存款；押金，保证金；存放物
	vt.	存放，堆积
	vi.	沉淀
dual	adj.	双的；二重的，双重的
inclinable	adj.	倾向于……的
reverse	n.	相反；背面，反面；倒退
	adj.	相反的；倒转的；颠倒的
	vt.	颠倒，倒转

课文导读 及 参考译文

（一）课文导读

在冲压生产中，使金属坯料产生塑性变形，形成一定角度或一定弯曲半径的零件的加工方法称为弯曲。弯曲成形可以使用模具在普通压力机上进行，也可以在专用的折弯机等设备上进行。在进行弯曲操作时，金属板料在小于其强度极限的拉应力和压应力作用下得到冲压，不发生明显的厚度改变。弯曲工序要注意回弹的影响，即已经被弯曲的金属仍然保留一些原来的弹性，在凸模移开之后，金属还可以恢复一些弹性。

（二）课文参考译文

弯曲成形

在用来完成冲裁工序的相同设备即曲柄、偏轴和凸轮压力机上，也可以进行弯曲和成形工序。在进行弯曲操作时，金属板料在小于其强度极限的拉应力和压应力作用下得到冲压，不发生明显的厚度改变。在压弯机中，简单的弯曲操作是指对金属板料的直角弯曲操作。其他的弯曲操作，如卷曲、接合和折叠，虽然程序稍微多一点，但都是类似的。弯曲力可以使用以下经验公式进行计算：

$$F = \frac{1.33 L S t^2}{W}$$

式中，

F——需要的弯曲力，单位为吨（兆帕）
L——弯曲线的长度，单位为英寸（毫米）
S——屈服应力，单位为吨/平方英寸（帕/毫米）
t——金属板厚度，单位为英寸（毫米）
W——V型或者U型凹模的宽度

V型或者U型模具是一种凹模和凸模配对。这种凹模和凸模通常在金属的板面形成一个90°的内角。1.33的经验恒定值是与金属厚度成比例的开模因素。

在设计一个用于弯曲的矩形截面时，必须确定允许弯曲的金属的量，因为板料在弯曲时，外层的纤维被拉长，内层的部分被压缩。在这项操作中，截面的中性轴向压缩的一面移动，这就使得更多的纤维得到拉伸。金属板料的厚度有细微的减少，压缩面的宽度加大，而另一面的宽度则减小。尽管弯曲的正确长度可由经验公式决定，但它还深受金属的物理特性的影响。已经被弯曲的金属仍然保留一些原来的弹性。如图10-1所示，在凸模移开之后，金属会有一些弹性回复。这种现象称作回弹。

图10-2所示为一个典型的应力应变曲线，这个曲线说明了弯曲操作中屈服点的重要性。如果A-B部分被伸展至A-C的长度，然后被释放，它将还原至原来的长度A-B。如果A-B部分在压力弯曲操作下被拉伸至比例极限L以上达到A-D的长度，那么当压力释放之后，它的最终长度是A-E。在这个例子中，回弹范围是E-D。直线M-E与L-B平行。因此，根据金属的应力应变特征和所受的载荷，就有可能确定回弹的范围。

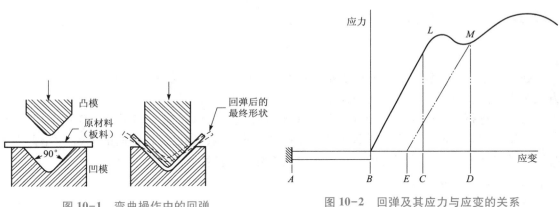

图10-1　弯曲操作中的回弹　　　　　图10-2　回弹及其应力与应变的关系

可以通过适当地加大弯曲量来纠正回弹，以保证压力释放后，零件能够恢复到其正确的形状。回弹在大弯曲半径操作中更加明显。最小弯曲半径因金属材料的延展性和厚度而异。

图10-3所示是用来将平钢带弯曲成U型件的成形模具。当凸模下压使零件成形时，推板被压下，致使在冲模底部的弹簧被压缩。凸模回升时，推板在弹力的作用下迫使工件离开模具。这种装置在大多数成形工序中都是必要的，因为金属会挤压冲模壁，带来工件卸料的困难。用于上冲程中的顶杆可以卸下容易粘住凸模的零件。

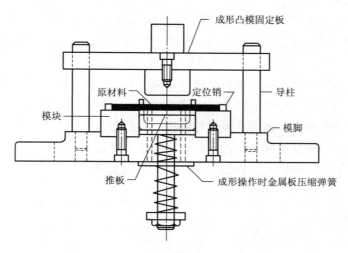

图 10-3 成形凸模与凹模

(三) 阅读材料 (5) 参考译文

组合、冲孔、弯曲之实例

图 10-4 所示为一个零件反置于模具上的实例；另外，在落料的同时还对其进行冲孔操作以提高产量。但是这种模具也面临同样的问题，即操作中会形成轻微的回弹并且也无法解决这种现象。然而如果这个问题没有重要的影响的话，这种模具设计就是很好的了。

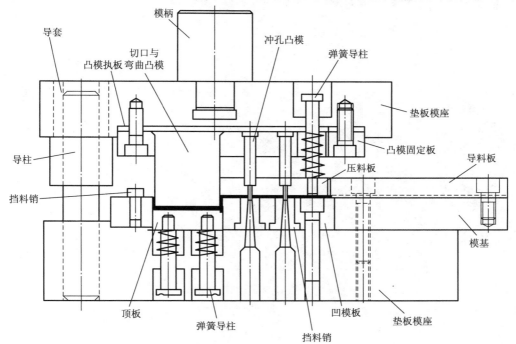

图 10-4 组合、冲孔、弯曲复合模具，为了保证这种生产工序的全面实行，就需要一个具备气喷净化装置以快速清洁表面的可倾式压力机

条料被传送推到挡料销 E，然后具有切口和弯曲双项功能的凸模下压处于落料凸模 N 和衬垫 G 之间的部分板料。接着模具继续运作将传送至凹模 D，对其进行如图所示的端部弯曲操作。

　　如果不采用某种形式的定位销定位条料，使两个凸模在原始坯料上冲出必需的孔的话，第一块坯料就会被浪费掉，但是孔会出现在随后的任一剪切部分，所以弯曲操作完成后的元件上就会出现位置正确的孔。条料在通过导料板 C 之后被压料板 P 平整地放在模具表面上。

　　冲压操作完成之后成品零件遗留在模具表面的这类模具最好安装在具备气压吹扫功能的可倾式压力机上来生产，以确保在压力机连杆再次下压之前，零件的表面是清洁的。

Lesson 11

Drawing Operation

Three bent flanges are shown in Fig. 11-1, The first one [Fig. 11-1(a)] is the simple straight bend. The stretch flange and shrink flange [Fig. 11-1(b) and (c), respectively] involve a plastic flow of metal that does not take place in a straight-bend flange. This plastic flow or adjustment of metal is characteristic of all drawing operations. Stresses are involved that exceed the elastic limit of the metal so as to permit the metal to conform to the punch. However, these stresses cannot exceed the ultimate strength without developing cracks. If the stretch flange [Fig. 11-1(b)] is considered to be a section of a circular depression that has been drawn, the metal in arc aa must have been stretched to a'a'. The action is a thinning one and must be uniform to avoid cracks. In the shrink flange (Fig. 11-1(c)) the action is just the opposite, and the metal in the flange is thickened.

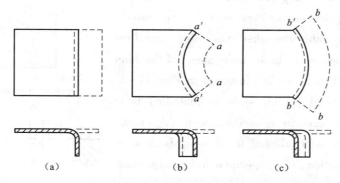

Fig. 11-1 Types of flanges

(a) Straight; (b) Stretch; (c) Shrink

Most drawn parts start with a flat plate of metal. As the punch is forced into the metal, severe tensile stresses are induced into the sheet being formed. At the same time the outer edges of the sheet that have not engaged the punch are in compression and undesirable wrinkles tend to form. This must be counteracted by a blank holder or pressure plate, which holds the flat plate firmly in place.

In a simple drawing operation of relatively thick plates the plate thickness may be sufficient to counteract wrinkling. This may be done in a single-acting press as shown in Fig. 11-2. Additional draws may be made on the cup-shaped part, each one elongating it and reducing the wall thickness.

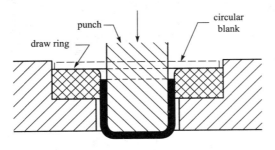

Fig. 11-2 Arrangement of punch and die for simple drawing operations

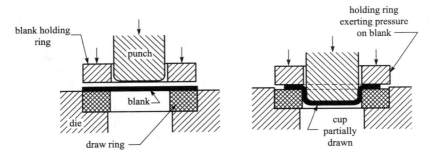

Fig. 11-3 Action of blankholder and punch in a drawing operation

Most drawing, involving the shaping of thin metal sheets, requires double-acting presses to hold the sheet in place as the drawing progresses. Presses of this type usually have two slides, one within the other. One slide controlling the blank-holding rings moves to the sheet ahead of the other to hold it in place. This action is illustrated in Fig. 11-3. The motion of the blank-holding slide is controlled by a toggle or cam mechanism in connection with the crank. Hydraulic presses are well adapted for drawing because of their relatively slow action, close speed control, and uniform pressure. Fig. 11-4 shows a sectional diagram of an inverted drawing die. The punch is stationary and is mounted on the bed of the press. As the die descends, the blank is contacted; then as its downward movement continues, the

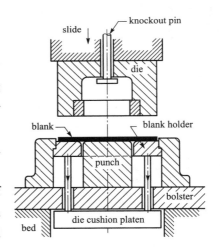

Fig. 11-4 Drawing operation using an inverted punch

blank-holding ring maintains contact with the blank during drawing. By the use of a die cushion to control the holder the pressures on he blank can be increased and controlled.

The pressure applied to the punch necessary to draw a shell is equal to the product of the cross-sectional area and the yield strength Y_s of the metal. A constant that covers the friction and bending is necessary in this relationship. The pressure P for a cylindrical shell may be expressed by the empirical equation

Chapter 3　Press Process and Die Design

$$P = \pi dtS(D/d - 0.6)$$

where

　　D—Blank diameter, in

　　d—Shell diameter, in

　　t—Metal thickness, in

　　S—Tensile strength, psi

The amount of clearance between a punch and die for blanking is determined by the thickness and kind of stock. For thin material the punch should be a close-sliding fit. For heavier stock the clearance must be larger to create the proper shearing action on the stock and to prolong the life of the punch.

There is a difference of opinion as to the method designating clearance. Some claim that clearance is the space between the punch and die on one side or one-half difference between the punch and die sizes. Others consider clearance as the total difference between the punch and die sizes. For example, if the die is round, clearance equals die diameter minus punch diameter. The advantage of designating clearance as the space on each side is particularly evident in the case of dies with irregular form or shape. Whether clearance is deducted from the diameter of the punch or added to the diameter of the die depends on the work. If a blank has a required size, the die is made smaller. When holes of a certain size are required, the punch is ground to the required diameter and the die is made larger.

 Words and Expressions

counteract [ˌkauntəˈrækt]	vt.	抵消，中和；阻碍
cushion [ˈkuʃən]	n.	垫子，软垫，衬垫
	v.	加衬垫
deduct [diˈdʌkt]	vt.	扣除，演绎
flange [flændʒ]	n.	边缘；轮缘；凸缘
	vt.	给……装凸缘
mount [maunt]	n.	衬纸；装配；装上，设置；安放
stationary [ˈsteiʃ(ə)nəri]	adj.	固定的
velour [vəˈluə]	n.	[纺] 丝绒，天鹅绒

1. This plastic flow or adjustment of metal is characteristic of all drawing operations.

译文：金属的这种塑性流动或调节是所有拉深工序的特征。

解析：be characteristic of：具有……的特征。

79

2. Stresses are involved that exceed the elastic limit of the metal so as to permit the metal to conform to the punch.

译文：所加的应力应超过金属的弹性极限，以便金属与凸模贴模。

解析：that exceed the elastic limit of the metal 是定语从句，修饰中心词 stresses。so as to：为了。conform to：符合；适合；与……一致。

3. Most drawing, involving the shaping of thin metal sheets, requires double-acting presses to hold the sheet in place as the drawing progresses.

译文：大多数涉及薄板成形的拉深，都需要双重的压力作用，以使板料在拉深过程中保持在合理的位置。

解析：句中的 involving the shaping of thin metal sheets 是现在分词短语作伴随状语。

4. By the use of a die cushion to control the holder the pressures on he blank can be increased and controlled.

译文：通过使用顶板来控制压料板，就可以增加和控制拉深模上的压力。

解析：By the use of a die cushion to control the holder 在句中作方式状语。

5. The pressure applied to the punch necessary to draw a shell is equal to the product of the cross-sectional area and the yield strength Y_s of the metal.

译文：拉深壳体时作用于凸模上的压力等于凸模的横截面面积与被拉深金属的屈服强度的乘积。

解析：句中 applied to the punch necessary to draw a shell 是过去分词短语作后置定语，修饰中心词 the pressure，其中 necessary to draw a shell 为形容词短语作后置定语，修饰中心词 punch。

6. Some claim that clearance is the space between the punch and die on one side or one-half difference between the punch and die sizes.

译文：一些人认为，凸模和凹模之间的间隙是由于两者之间的尺寸不一样而形成的凸、凹模之间的单边间隙或双边空隙。

解析：句中的 clearance is the space between the punch and die on one side or one-half difference between the punch and die sizes 是谓语 claim 的宾语从句。

7. Others consider clearance as the total difference between the punch and die sizes. For example, if the die is round, clearance equals die diameter minus punch diameter.

译文：其他人则把间隙看做是凹模尺寸和凸模尺寸之间总的差值。例如，如果凹模是圆的，间隙就等于凹模的直径减去凸模的直径。

解析：consider...as...：把……看做……。

8. The advantage of designating clearance as the space on each side is particularly evident in the case of dies with irregular form or shape.

译文：在模具的形式或形状不规则的情况下，把单边间隙看做是间隙的好处是尤其显然的。

解析：designate...as...：把……看做……。in case of：在……情况下。

Chapter 3   Press Process and Die Design

  Exercise

Fill in the blanks with the proper words.

1. The stretch flange and shrink flange _____ a plastic flow of metal that does not _____ in a straight-bend flange.
2. If the stretch flange is considered _____ be a section _____ a circular depression that has been drawn, the metal _____ arc aa must have been stretched to a′a′.
3. Most drawn parts start _____ a flat plate _____ metal. _____ the punch is forced _____ the metal, severe tensile stresses are induced _____ the sheet being formed.
4. Hydraulic presses are well _____ _____ drawing because _____ their relatively slow action, close speed control, and uniform pressure.
5. _____ heavier stock the clearance must be larger to _____ the proper shearing action _____ the stock and to _____ the life of the punch.

课文导读 及 参考译文

（一）课文导读

拉深是利用模具将平板坯料拉制成开口空心零件或将开口空心零件加工成其他形状的空心零件的一种冲压加工方法。拉深也称为拉延。拉深是通过拉伸力使金属坯料发生延伸或收缩。典型的拉深件是圆筒形（或杯形）零件。

（二）参考译文

拉　深

图 11-1 所示为 3 种弯曲形状。图 11-1(a)是一种简单的直弯变形。而图 11-1(b)、图 11-1(c)分别为外缘伸展（翻边）和外缘收缩（翻边），这两者都意味着金属产生了塑性流动，而直弯凸缘不会发生这种流动。金属的这种塑性流动或调节是所有拉深工序的特征。所加的应力应超过金属的弹性极限，以便金属与凸模贴模。但是所加的应力不能超过其强度极限以避免形成裂缝。如果把外缘伸展当作是已经拉深成形的圆形凹陷面的一部分，那么图中弧 aa 部分的金属就必须被伸展至弧 a′a′。这种行为是一种变薄工序，在操作中同样也必须避免裂缝。在外缘收缩[图 11-1(c)]操作中，动作刚好相反，外缘收缩的金属被增厚了。

多数拉深零件的原形是平板金属坯料。凸模进入金属坯料时，将在板料上形成剧烈的拉伸应力，同时没有受到凸模冲压的板料外壁受到收缩，并且往往会形成不必要的褶皱。必须通过用来保持平板坯稳定在合理位置的压料板或压力板来抵制这种现象的产生。

在对相对较厚的金属板进行简单拉深时，金属板的厚度也许足以抵制褶皱的产生。这种情况可能会出现在如图 11-2 所示的单工序冲压过程中。其他的拉深可以按圆筒形进行，每种拉深都可以使板料延伸并减少金属厚度。

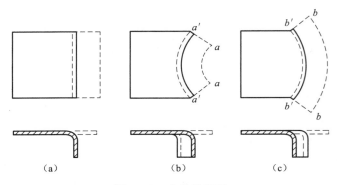

图 11-1　凸缘的类型

（a）直弯；（b）外缘伸展（翻边）；（c）外缘收缩（翻边）

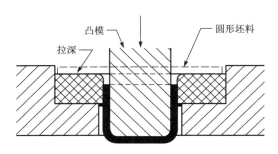

图 11-2　简单拉深操作的凸模和凹模的布置

  大多数涉及薄板成形的拉深，都需要双重的压力作用，以使板料在拉深过程中保持在合理的位置。这种类型的压力机通常有两个滑块，一个滑块嵌在另一个滑块之内。用来控制坯料压料板的一个滑块在另一个滑块之前移向板料使其保持在合理的位置。图 11-3 显示的就是这种操作。压料的滑块的运动由一个曲柄或者凸轮机构连同曲柄轴控制。液压压力机由于其运作缓慢、精密的速度控制和均衡压力的特点，非常适合于进行拉深。图 11-4 是一个倒装的拉深模截面图。凸模是固定的，安置在压力机的工作台上。凹模下降时与坯料接触，随着凹模的不断下降，压料板在拉深过程中保持与坯料相接触。通过使用顶板来控制压料板就可以增加和控制拉深模上的压力。

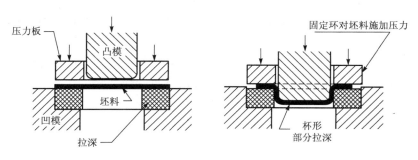

图 11-3　压料板的作用以及拉深操作中的凸模

  拉深壳体时作用于凸模上的压力等于凸模的横截面积与被拉深金属的屈服强度的乘积。在这个关系式中也必然存在一个摩擦与弯曲恒定值。拉深一个圆筒形件的压力可以用经验方程式表示：

$$P = \pi dtS(Dld-0.6)$$

式中，

D——坯料直径，单位：英寸

d——金属壳直径，单位：英寸

t——金属板厚度，单位：英寸

s——强度极限，单位：帕

用来落料的凹模与凸模之间的间隙取决于原料的厚度与种类。对于薄板料，凸模应是小间隙。对于厚板料而言，就必须取更大的间隙以便在原料上完成正确的冲裁操作，并且延长凸模的使用寿命。

人们对于设定间隙值的方法有着不同的意见。一些人认为，凸模和凹模之间的间隙是由于两者之间的尺寸不一样而形成的凸、凹模之间的单边间隙或双边间隙。其他人则把间隙看做是凹模尺寸与凸模尺寸之间总的差值。例如，如果凹模是圆的，间隙就等于凹模的直径减去凸模的直径。在模具的形式或形状不规则的情况下，把单边间隙看做是间隙的好处是尤其显然的。间隙究竟是从凸模直径上扣除还是加在凹模直径上，取决于工序的性质。如果落料件有要求的尺寸，凹模就做小些。当冲孔尺寸有要求时，凸模以所要求的尺寸为基准，凹模做大些。

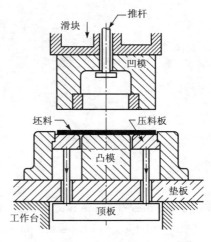

图 11-4 使用倒装凸模的拉深工序

Lesson 12

Compound and Progressive Dies

Compound dies combine two or more operations at one station as illustrated in Fig. 12-1. Strip stock is fed to the die, two holes are punched, and the piece is blanked on each stroke of the ram. When the operations are not similar, as in the case of a blanking and forming operation, dies of this type are frequently known as combination dies.

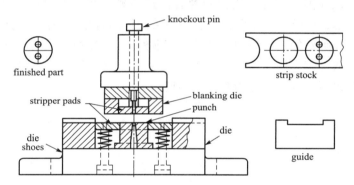

Fig. 12-1 Compound punch and die

A progressive die set performs two or more operations simultaneously but at different stations. A punch-and-die set of this type is shown in Fig. 12-2. As the strip enters the die, the small square hole is punched. The stock is then advanced to the next station, where it is positioned by the pilot as the blanking punch descends to complete the part.

This general type of design is simpler than the compound dies, because the respective operations are not crowded, together. Regardless of the number of operations to be performed, the finished part is not separated from the strip until the last operation.

In the perforation of metal one or two rows of punches are employed. The metal sheet is moved incrementally through the press each time the punches are withdrawn. The punched holes may be almost any shape. All nonbrittle metals can be perforated.

Misalignment of punch and die causes excessive pressures, shearing or chipping of die edges, or actual breaking of the tools. Such action may occur through shifting even though the setup was originally correct. To prevent these occurrences, proper alignment is ensured by guide rods at two or four corners of the die that fit into holes in the punch holders. These dies are known as pillar

Chapter 3 Press Process and Die Design

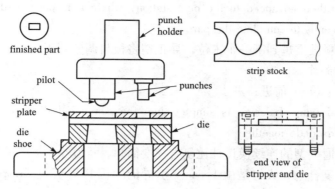

Fig. 12-2 Progressive punch and die

dies. This arrangement of having the punch and die held in proper alignment facilitates the setting up of the tools. A similar arrangement, known as a subpress die and occasionally used on small work, employs a punch and die mounted in a small frame so that accurate alignment is always maintained. Pressure is applied by a plunger that extends out of the top of the assembly.

Words and Expressions

brittle ['britl]	adj.	易碎的，脆弱的
crowd [kraud]	n.	人群，群众；一堆(东西)；一伙
	v.	群集，拥挤，挤满
incremental [inkri'mentəl]	adj.	增加的
misalignment ['misəlainmənt]	n.	未对准
occasionally [ə'keiʒənəli]	adv.	有时候，偶尔
occurrence [ə'kʌrəns]	n.	发生，出现；事件，发生的事情
perforation [pɜːfə'reiʃ(ə)n]	n..	穿孔
pillar ['pilə]	n.	[建]柱子，栋梁
pilot ['pailət]	n.	导杆；控制器
	vt.	指引，引导

Notes

1. When the operations are not similar, as in the case of a blanking and forming operation, dies of this type are frequently known as combination dies.

译文：当冲压的工序不同时，例如，能够在一副模具上完成落料和成形，这种模具通常称之为组合模。

解析：be known as：被称之为……。

2. The stock is then advanced to the next station, where it is positioned by the pilot as the blanking punch descends to complete the part.

译文：然后板料被推进到下一个工位，导正销将板料调整到正确位置上，落料凸模下行冲裁，完成零件的制作。

解析：advance：（v.）前进；进步。

3. This general type of design is simpler than the compound dies, because the respective operations are not crowded, together.

译文：这种普通类型的连续模设计比复合模要简单，因为各个工序是分开完成的。

解析：此句为复合句，从句是由 because 引导的原因状语从句。respective：各自的，分别的。

4. Regardless of the number of operations to be performed, the finished part is not separated from the strip until the last operation.

译文：无论前面完成了多少工序，完成后的零件直到最后的工序才从条料上分离下来。

解析：regardless of：不管，不顾。

5. Such action may occur through shifting even though the setup is originally correct.

译文：这种情况可能发生在装置的移动过程中，即使装置最初并未出错。

解析：even though：即使（引导让步状语从句）。occur：发生，出现。

6. To prevent these occurrences, proper alignment is ensured by guide rods at two or four corners of the die that fit into holes in the punch holders.

译文：为了防止这种未对正现象发生，可以通过在下模座上的两个角或四个角安装与上模座上的导向孔相配合的导柱来保证上、下模的正确对正。

解析：句中的 to prevent these occurrences 为不定式作目的状语。that fit into holes in the punch holders 是定语从句，修饰中心词 guide rods。

7. This arrangement of having the punch and die held in proper alignment facilitates the setting up of the tools.

译文：以适当的排列形式将凹模与凸模安置在一起的排列方法更加便利于模具的装配。

解析：facilitate：使便利，使容易。

8. A similar arrangement, known as a subpress die and occasionally used on small work, employs a punch and die mounted in a small frame so that accurate alignment is always maintained.

译文：常见的、偶尔用于小规模生产的次压模的排列是一种类似的情况，它是把凹模和凸模置于一个小型框架中以经常保持对正的准确性。

解析：句中的 known as a subpress die and occasionally used on small work 为过去分词短语作后置定语，修饰 a similar arrangement。

9. Pressure is applied by a plunger that extends out of the top of the assembly.

译文：由模具顶端向外伸出的模柄传递冲压力。

解析：apply：应用，把……应用于，投入或应用于一项特定用途。

Chapter 3 Press Process and Die Design

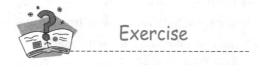   Exercise

Fill in the blanks with the proper words.

1. Strip stock is fed _____ the die, two holes are punched, and the piece is blanked _____ each stroke _____ the ram.
2. A progressive die set performs two or more operations _____ but at different stations.
3. The stock is then _____ _____ the next station, where it is positioned _____ the pilot as the blanking punch descends _____ complete the part.
4. The metal sheet is moved _____ through the press each time the punches are _____ .
5. _____ of punch and die causes _____ pressures, shearing or chipping of die edges, or actual breaking of the tools.

Reading Materials (6)

Combination and Compound Dies

Confusion sometimes arises regarding combination and compound tools largely because the names are similar, but they differ considerably in their design and application. In order to appreciate the function of a combination tool, any newcomer to press tools must know that this type of equipment is designed to perform two operations on a strip or precut piece of material simultaneously. There are obviously several such "combinations" and the most well known are: blank and pierce, blank and cup, blank and plunge, and blank, draw and trim.

A washer-producing tool is a suitable example to use as the operations of blanking the periphery and piercing the central hole are carried out simultaneously. The action differs from a follow-on tool in which the hole is punched and then the strip is fed further through the tool to another punch that removes the partially-completed blank from the material. In a combination tool both of the cutting stages are carried out without moving the strip.

A typical tool for producing washers is shown in Fig. 12-3, and the first significant change from the follow-on tool is the absence of a material guide, and although a guide can easily be arranged on each side of the die, a strip feeding mechanism outside the tool is usually employed. However, this does not affect the design of the tool in which, as shown, the base locates the punch-cum-blanking die A, which has a dual function to perform because the outside diameter serves as the piercing punch for the outside diameter of the washer, while the bore is the die when the centre piece of scrap material is removed. The blanking die is thus carried in the upper member with a central punch and in a recess machined behind this is a headed ejector rod that operates four pins used for moving the actual ejector G. Sometimes this ejector is spring-loaded, but a mechanical ejector of the type shown is better for simple components because the part leaves the strip and by using an air blast it is blown from the die surface. With the spring type of ejector the blank is reinserted into the strip and carried along and will require removing by hand. Meanwhile the centre piece of scrap falls away through the bolster.

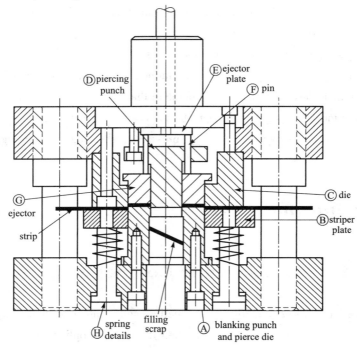

Fig. 12-3 A combination tool for producing a washer-type component where both speed and concentricity of the diameters is essential. This idea is applicable for details of all sizes, the only restriction being the dimensions of the hole in the press bed to allow a scrap disc to fall through

A combination tool is not restricted to pressing circular details of the type depicted here. One further important factor is that producing a single washer in the manner shown is the most uneconomical way of carrying out the task because of the high material wastage. At least three or preferably five washers can be pressed simultaneously from the strip in order to reduce wastage to a minimum. This practice is, of course, only feasible for small washers.

Words and Expressions

bore	n.	枪膛；孔
	v.	钻孔
ejector	n.	驱逐者；排出器

newcomer	n.	新来者；新到的移民
precut	vt.	按规格裁切
rod	n.	杆，棒
simultaneously	adv.	同时地
uneconomical	adj.	不经济的，浪费的

(一) 课文导读

复合模是在一个工位上同时完成2个或2个以上工序。连续模是同时完成2个或更多的工序，但是是在不同的工位上。连续模又称多工位级进模。在复合模与连续模操作中，凸模与凹模没有对齐就会产生过大的压力，从而会剪切或碎裂模具边缘或者造成对模具的实际损害，所以在操作中要避免此种现象的产生。

(二) 课文参考译文

复合模与连续模

图12-1所示的复合模是指在一个工位上同时执行2个或2个以上工序。条料被送至模具，滑块每次冲压形成两个孔，并将零件落料。当冲压的工序不同时，例如，在一副模具上完成落料和成形，这种模具通常称之为组合模。

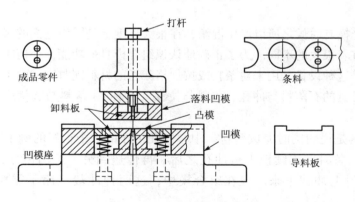

图12-1 复合凸模与凹模

一副连续模同时完成两个或更多工序，但是是在不同的工位上。图12-2所示就是这种类型的凸模与凹模。条料被送进模具内，先冲压形成小的方形孔。然后板料被推进到下一个工位，导正销将板料调整到合理位置上，落料凸模下行冲裁，完成零件的制作。

这种普通类型的连续模设计比复合模要简单，因为各个工序是分开完成的。无论前面完成了多少工序，完成后的零件直到最后的工序才从条料上分离下来。

金属板料的冲孔需要采用一排或两排凸模。每次凸模回升时，金属板料不断被移动通过压力机。可以冲压形成几乎任何形状的孔。所有不易碎的金属都能被冲孔。

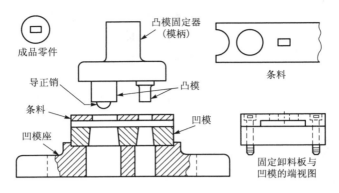

图 12-2　连续凸模与凹模

凸模与凹模没有对正就会产生过大的压力，从而会剪切或碎裂模具边缘或者造成对模具的实际损害。这种情况可能发生在装置的移动过程中，即使装置最初并未出错。为了防止这种未对正现象发生，可以通过在下模座上的两个或四个角上安装与上模座上的导向孔相配合的导柱来保证上、下模的正确对正。这些模具称作导柱模。以适当的排列形式将凹模与凸模安置在一起的排列方法更加便利于模具的装配。常见的、偶尔用于小规模生产的次压模的排列是一种类似的情况，它是把凹模与凸模置于一个小型框架中以经常保持对正的准确性。由模具顶端向外伸出的模柄传递冲压力。

（三）阅读材料（6）参考译文

组合模与复合模

组合模与复合模有时候之所以产生混淆，在很大程度上是因为它们的名字类似，但它们在设计与应用上有着很大的不同。为了正确地认识组合模具的功能，任何刚进入模具行业的新手都必须知道，这种设备是用来对条料或按规格裁切的材料同时完成两个工序的。有许多这样的组合，最常见的有落料与冲孔、落料与拉深（冲杯）、落料与起伏成形、冲孔、拉深和剪边。

垫圈生产模具是说明同时完成落料、中心冲孔工序的一个很好的例子。此类模具的冲压不同于连续模。条料在连续模上被冲孔之后，再传送到另一个凸模处，由其将已部分完成的工件从板料上冲裁下来。但在组合模中，这两种冲裁工序不需要移动条料就能完成。

图 12-3 为一套典型的垫圈生产模具，组合模与连续模的一个很大不同是，它没有使用导料板，虽然在模具的每个面上可以很容易地装上导料板，但是组合模通常采用一种位于模具外侧的送料机构送料。然而这并没有影响到这种模具的设计。如图所示，模座安装有凸凹模 A，这种模具要完成双重工序，因为其外径用作垫圈外径的冲裁凸模，同时其内孔又是切除中心废料部分的凹模。落料凹模与一个中心凸模一起安装在上模，在其凹进处还安装有一个推杆，它控制四个顶杆来移动推料块 G。有的推料块还装有弹簧，但图中所示的机械推料的模具更适用于冲压小型工件，因为零件从条料脱离后可用空气吹扫出模具表面。在配备弹簧推料块的模具中，坯料被重新落入条料并被随料运送，所以还要求手工移动，其间中心部

分的废料穿过垫板而脱落。

组合模的功能并不只局限于冲压以上所描述的圆形零件。一个更重要的因素是按图所示的方式生产单一的垫圈是最不经济的,因为它会造成大量的材料浪费。在所产生的废料上至少还可以同时冲压出 3 个或 5 个垫圈,以将浪费减至最低限度。当然这种做法只适合小型垫圈的生产。

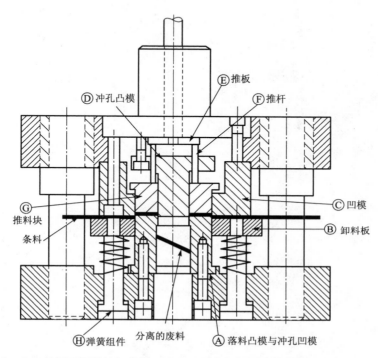

图 12-3　生产垫圈型零件的组合模,速度以及直径同心是其基本要求。这种方法适合于所有规格的零件,唯一的限制就是压力机工作台的废料脱落口的尺寸

Chapter 4

Plastics Forming and Mold Design
（塑料成型工艺及模具）

Lesson 13

Summary of Plastics

Plastics are an important class of materials with an extremely wide range of mechanical, physical, and chemical properties. The first plastics (natural polymers) were produced in the 1860s; however, modern plastics technology (synthetic polymers) began in the 1920s. These materials are called plastics, from the Greek plastics meaning they can be molded and shaped.

Plastics are characterized by the following properties: low density, low strength and elastic modulus, low thermal and electrical conductivity, high chemical resistance, and high coefficient of thermal expansion. They can be cast, formed, machined, and joined into different shapes and are available in a wide variety of properties, colors, and opacities. Also included in this group of materials are rubbers and elastomers.

Because of their many unique properties, plastics have increasingly replaced metallic components in numerous applications. This trend is explained by considerations with respect to service requirements, strength-to-weight ratio, design and cost of the material, and ease of manufacturing.

With various additives and reinforcements, plastics are used in a wide range of consumer and industrial products. Some of the major applications are in automotive, electrical and electronic products such as integrated circuits, mechanical equipment, food and beverage containers, packaging, signs, housewares, textiles, safety equipment, toys, appliances, and optical equipment.

 Words and Expressions

additive [ˈæditiv]	adj.	附加的，加成的，添加的；加法的
	n.	添加剂
available [əˈveiləbl]	adj.	可用到的；可利用的；有用的；有空的
beverage [ˈbevəridʒ]	n.	饮料
cast [kɑːst]	n.	投掷；铸件；脱落物，
	v.	投，抛；投射；浇铸
coefficient [kəuiˈfiʃənt]	n.	[数]系数

component [kəm'pəunənt]	n.	成分
	adj.	组成的，构成的
container [kən'teinə]	n.	容器(箱，盆，罐，壶，桶，坛子)；集装箱
density ['densiti]	n.	密度
elastic modulus		弹性模量
elastomer [i'læstəmə(r)]	n.	弹性体；人造橡胶
houseware ['hausweəz]	n.	家用器皿
integrated circuit		集成电路
metallic [mi'tælik]	adj.	金属(性)的
opacity [əu'pæsiti]	n.	不透明性
optical ['ɔptikəl]	adj.	眼的，视力的；光学的
packaging ['pækidʒiŋ]	n.	包装
polymer ['pɔlimə]	n.	聚合体
reinforcement [ˌriːin'fɔːsmənt]	n.	增援；加强；加固；援军
strength [streŋθ]	n.	力，力量，力气；实力；兵力；浓度
synthetic [sin'θetik]	adj.	合成的；人造的；综合的
textile ['tekstail]	n.	纺织品
	adj.	纺织的
thermal conductivity		导热性[系数]

Notes

1. Plastics are an important class of materials with an extremely wide range of mechanical, physical, and chemical properties

译文：塑料是一种具有非常广泛的机械、物理和化学性能的重要材料。

2. The first plastics (natural polymers) were produced in the 1860s; however, modern plastics technology (synthetic polymers) began in the 1920s.

译文：最初的塑料（天然聚合物）生产于19世纪60年代；但是，现代塑料技术（合成聚合物）出现于20世纪20年代。

解析：此句中 be produced in 译为"生产于"。be produced from 译为"用……制造"。

3. Plastics are characterized by the following properties: low density, low strength and elastic modulus, low thermal and electrical conductivity, high chemical resistance, and high coefficient of thermal expansion.

译文：塑料具有以下特点：低密度、低强度和弹性模量、低导热性和低导电性、高抗化学阻力和高热膨胀系数。

4. They can be cast, formed, machined, and joined into different shapes and are available in a wide variety of properties, colors, and opacities.

译文：塑料能浇铸、成型、机械加工和连成不同的形状，并且可以具有不同的性能、颜色和光透性。

5. Because of their many unique properties, plastics have increasingly replaced metallic components in numerous applications.

译文：因为具有许多独特的性质，塑料在很多应用中已经逐渐地取代了金属成分。

6. This trend is explained by considerations with respect to service requirements, strength-to-weight ratio, design and cost of the material, and ease of manufacturing.

译文：这种趋势可以从以下这些方面进行解释，即产品的服务需求、强度和自重的比例、材料的设计和成本以及制造的简易性。

7. With various additives and reinforcements, plastics are used in a wide range of consumer and industrial products.

译文：塑料含有各种添加剂和增强相，广泛地应用于生活用品和工业制品中。

8. Some of the major applications are in automotive, electrical and electronic products such as integrated circuits, mechanical equipment, food and beverage containers, packaging, signs, housewares, textiles, safety equipment, toys, appliances, and optical equipment.

译文：塑料主要应用于汽车产品、电气和电子产品中，例如集成电路、机械装备、食品和饮料罐、包装、标记、家用器皿、纺织品、安全装置、玩具、器具和光学设备。

解析：此句中 some of the major applications are in… 译为"塑料主要应用于……"。

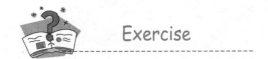

Fill in the blanks according to the text.

1. The first plastics (natural polymers) were produced in the 1860s; _____, modern plastics technology (synthetic polymers) began in the 1920s.
2. These materials are called plastics, from the Greek plastics meaning they can be molded and _____.
3. _____ their many unique properties, plastics have increasingly replaced metallic components in numerous applications.
4. This _____ is explained by considerations with respect to service requirements, strength-to-weight ratio, design and cost of the material, and ease of manufacturing.
5. With various additives and reinforcements, plastics are used in a wide range of consumer and _____ products.

课文导读及参考译文

（一）课文导读

本文主要介绍了塑料的特性和用途。塑料具有以下特点：低密度、低强度和弹性模量、低导热性和低导电性、高抗化学阻力和高热膨胀系数。塑料主要应用于一些汽车产品、电气和电子产品中。

（二）课文参考译文

塑料的概述

　　塑料是一种具有非常广泛的机械、物理和化学性能的重要材料。最初的塑料（天然聚合物）产生于19世纪60年代；但是，现代塑料技术（合成聚合物）出现于20世纪20年代。希腊语中的"塑料"一词意指"能被模压和成型"，因此这些材料被称为塑料。

　　塑料具有以下特点：低密度、低强度和弹性模量、低导热性和低导电性、高抗化学阻力和高热膨胀系数。塑料能浇铸、成型、机械加工和连成不同的形状，并且具有不同的性能、颜色和光透性。该类材料中也包括橡胶和弹性体。

　　因为具有许多独特的性质，塑料在很多应用中已经逐渐地取代了金属成分。这种趋势可以从以下这些方面进行解释：即产品的服务需求、强度和自重的比例、材料的设计和成本以及制造的简易性。

　　塑料含有各种添加剂和增强相，广泛地应用于生活用品和工业制品中。塑料主要应用于一些汽车产品、电气和电子产品中，例如集成电路、机械装备、食品和饮料罐、包装、标记、家用器皿、纺织品、安全装置、玩具、器具和光学设备。

Lesson 14

The Structure of Plastics

Plastics are composed of polymer molecules and various additives. Polymers are long-chain molecules (also called giant molecules or macromolecules), which are formed by polymerization; that is, linking and cross-linking of different monomers.

Monomers

A monomer is the basic building block of polymers. The word mer (from the Greek meros, meaning part) indicates the smallest repetitive unit, similar to the term unit cell in crystal structures. Thus, polymer means many mers or units. Monomers are organic materials. They include carbon atoms joined in covalent bonds (electron sharing) with other atoms such as hydrogen, oxygen, nitrogen, fluorine, chlorine, silicon, and sulfur.

A typical monomer is the ethylene molecule shown in Fig. 14-1 (a). These molecules can be made to attach themselves to other ethylene molecules by a chemical reaction involving heat, pressure, and a catalyst. In this reaction, the double bonds between the carbon atoms open and the molecules arrange themselves in a long line [Fig. 14-1 (b)].

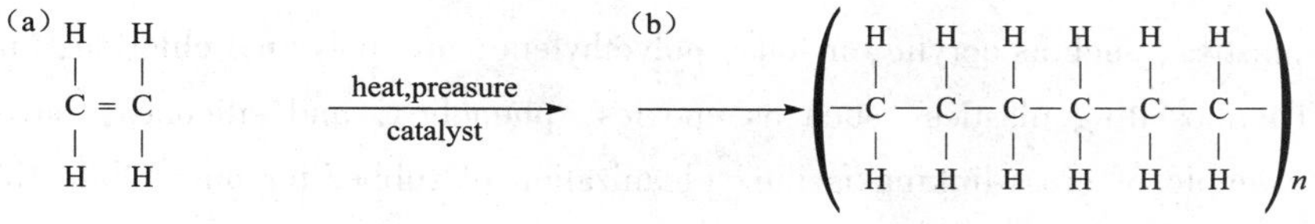

Fig. 14-1 Basic structure of polymer molecules
(a) Ethylene molecule; (b) Polyethylene molecule. Polymers

The polymer shown in Fig. 14-1 (b) is called a linear polymer because of its linear structure. The number of molecules in this chain (length of chain) is known as the molecular weight. A linear molecule does not mean that it is straight. These molecules (chain) are of different lengths and their arrangement is amorphous (without any long-range order). This arrangement is often described as a bowl of spaghetti, or worms in a bucket, all intertwined with each other. Because of the differences in the length of the chains, the molecular weight of a polymer is determined on a statistical basis.

Bonding

Although within each long-chain molecule there is covalent bonding (primary bonds), no such intermolecular bonding exists between different chains. The bonds between different chains (intramolecular), and between the overlapping portions of the same chain, are known as secondary bonds. Secondary bonds are van der Waals bonds, hydrogen bonds, and ionic bonds. Secondary bonds are all much weaker than the covalent bonds within the chain.

The difference in strength between the two bonds is on the order of one to two orders of magnitude. This difference is important, because it is the weaker secondary bonds that determine the overall strength of the polymer. (Thus, for instance, an object is only as strong as its weakest part.)

If the repeating units in a chain are all of the same type, the macromolecule is known as a homopolymer. However, in order to obtain certain special properties, two or three different types of monomers can be combined in a polymer. These are known as copolymers and terpolymers, respectively.

Branching and Linking

The properties of a copolymer depend not only on the two monomers, but also on their arrangement in the molecular structure (Fig. 14-2). Branching, for instance, interferes with the relative movement of the molecules and affects the resistance to deformation. Another kind of chain is cross-linking [Fig. 14-2 (c)], which is spatial (three-dimensional) network structure with adjacent chains tied together. Cross-linking has great influence on the properties of the polymer, such as in the vulcanization of rubber.

Fig. 14-2 Schematic illustration of polymer chains
(a) Linear; (b) Branched; (c) Cross-linked

Thermoplastics, such as acrylic, nylons, polyethylene, and polyvinyl chloride, are linear-chain molecules. Thermosetting plastics, such as epoxies, phenolics, and silicones, have cross-linked chains. An example of cross-linking is the vulcanization of rubber for automobile tires, each tire being one giant molecule.

Plastics made of linear chain structures are called thermoplastics, and those made of cross-linked chain are called thermosetting plastics, or simply, thermosets.

 Words and Expressions

adjacent [ə'dʒeisənt]   adj.   邻近的，接近的
bowl [bəul]   n.   碗；碗状物；木球
     v.   滚；(板球)；投球

catalyst ['kætəlist]   n.   催化剂

copolymer [kəu'pɔlimə]	n.	[化]共聚物
covalent [kəu'veilənt]	adj.	[化]共有原子价的,共价的
ethylene ['eθili:n]	n.	[化]乙烯,乙烯基
homopolymer [ˌhəumə'pɔlimə]	n.	[化]均聚(合)物
intermolecular[ˌintə(:)mə'lekjulə]	adj.	[化]分子间的存在(或作用)于分子间的
intertwine [ˌintə(:)'twain]	v.	(使)纠缠;(使)缠绕
monomer ['mɔnəmə]	n.	单体
overlapping ['əuvə'læpiŋ]		重叠;搭接
spaghetti [spə'geti]	n.	意大利式细面条
spatial ['speiʃəl]	adj.	空间的
statistical [stə'tistikəl]	adj.	统计的;统计学的
thermoplast [ˌθə:mə'plæstiks]	n.	热塑性塑料
thermosetting [ˌθə:məu'setiŋ]	adj.	热硬化性的
vulcanization [ˌvʌlkənai'zeiʃən]	n.	(橡胶的)硫化(过程),硫化

Notes

1. Plastics are composed of polymer molecules and various additives.

译文:塑料由聚合分子和各种添加剂组成。

解析:此句中 are composed of 译为"由……组成"。

2. Polymers are long-chain molecules (also called giant molecules or macromolecules), which are formed by polymerization; that is, linking and cross-linking of different monomers.

译文:聚合体是长链分子(也称巨大分子或高分子),它是通过聚合形成的,即单体的结合和交联。

解析:此句中 which 在句中引导非限制性定语从句。that is 的意思是"即,就是,换句话说,就是说,更确切地说",也可以说成 that is to say。

3. The word mer (from the Greek meros, meaning part) indicates the smallest repetitive unit, similar to the term unit cell in crystal structures.

译文:mer 这个词(来自希腊语的 meros,意思是"部分")表示最小的重复个体,与晶体结构中的术语"单位晶格"相似。

解析:句中 unit cell 的意思是"晶胞,单胞,格子单位,单位晶格,单位粒子"。be similar to:"与……相似"。

4. These molecules can be made to attach themselves to other ethylene molecules by a chemical reaction involving heat, pressure, and a catalyst.

译文:这些分子可以通过加热、加压和添加催化剂的化学反应过程,把它们自己依附在其他乙烯分子上。

解析:句中 attach to 的意思是"使依附,把……放在"。

5. A linear molecule does not mean that it is straight.

译文：线型分子并不意味着它是直的。

解析：此句中 mean 译为"意味"。

6. Although within each long-chain molecule there is covalent bonding (primary bonds), no such intermolecular bonding exists between different chains. The bonds between different chains (intramolecular), and between the overlapping portions of the same chain, are known as secondary bonds.

译文：虽然在每个长链分子之间有一个共价键（主键），但是在不同的链之间则不存在这种分子间的共价键。在相同链之间的重叠部分存在的键被称为次键。

7. If the repeating units in a chain are all of the same type, the macromolecule is known as a homopolymer.

译文：如果一条分子链中的重复单元都是同一类型，那么这种高分子称为均聚物。

8. The properties of a copolymer depend not only on the two monomers, but also on their arrangement in the molecular structure (Fig. 4.2).

译文：共聚物的性能不仅决定于两个单体，也取决于它们在分子结构中的排列（图14-2）。

解析：此句中 depend... on... 译为"决定于，被……决定，以……为条件，视……而定"。

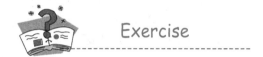

Exercise

Fill in the blanks according to the text.

1. Plastics are composed _____ polymer molecules and various additives.
2. The number of molecules in this chain (length of chain) is known as the _____.
3. The bonds between different chains (intramolecular), and between the overlapping portions of the same chain, are known as _____.
4. This difference is important, _____ it is the weaker secondary bonds that determine the overall strength of the polymer.
5. The properties of a copolymer depend not only on the two monomers, _____ on their arrangement in the molecular structure.

Reading Materials (7)

Additives

In order to impart certain specific properties, polymers are generally compounded with additives. Additives modify and improve certain characteristics such as stiffness, strength, color, weatherability, flammability, arc resistance for electrical applications, and ease of subsequent processing. These additives are described below.

Fillers

These materials are generally wood flour, silica flour, various minerals, powdered mica, and short fibers of cellulose, glass, and asbestos. Depending on their type, fillers improve the strength, hardness, toughness, abrasion resistance, and stiffness of plastics. These properties are maximized at various percentages of different types of polymer/filler combinations.

The effectiveness of a filler depends on the nature of the bond between the filler material and the polymer chains. Because of their lower cost, fillers also reduce the overall cost per unit weight of the polymer. Most thermoplastics and some thermosetting plastics contain fillers.

Plasticizers

These are added to give flexibility and softness to the polymer by lowering the glass-transition temperature. Plasticizers are low-molecular-weight solvents (with high boiling points, i.e., nonvolatile) that reduce the strength of the secondary bonds between the long-chain molecules of the polymer, thus making it soft and flexible. The most common application is in polyvinyl chloride (PCV), which remains flexible during its many uses.

Stabilizers

Most polymers are adversely affected by ultraviolet radiation and oxygen which weaken and break the primary bonds of the long-chain molecules. The polymer then becomes rigid and brittle. This is known as degradation. A typical example of protection against ultraviolet radiation is the compounding of rubber with carbon black (soot). The carbon black absorbs a high percentage of the ultraviolet radiation. Protection against degradation by oxidation, particularly at elevated temperatures, is done by adding antioxidants to the polymer. Various coatings are another means of protection against degradation.

Colorants

The great variety of colors available in plastics is obtained by the addition of colorants. These are either organic (dyes) or inorganic (pigments). The selection of a colorant depends on service temperature and exposure to light. Pigments, which are dispersed particles, generally have greater resistance to temperature and light than dyes do.

Flame Retardants

If the temperature is sufficiently high a polymer will ignite. The flammability (the ability to support combustion) of polymers varies considerably, depending on their composition, such as the chlorine and fluorine content. Polymethylmethacrylate, for example, continues to burn when ignited, whereas polycarbonate extinguishes itself. The flammability of polymers can be reduced either by making them from less flammable raw materials or by adding flame retardants to the compound. Common flame retardants are chlorine, bromine, and phosphorus compounds.

Lubricants

To reduce friction during subsequent processing and use and to prevent sticking to the molds, lubricants are added to polymers. Lubrication is also important in preventing thin polymer films from sticking together.

Polyblends

To give them a rubbery behavior, polymers can be blended with small amounts of rubbery polymers. These are finely dispersed throughout the polymer and improve its impact strength. These polymers are known as rubber-modified.

Words and Expressions

antioxidants		抗氧化(作用)，氧化抑制(作用)
asbestos	n.	[矿]石棉
cellulose	n.	纤维素
chloride	n.	[化]氯化物
colorant	n.	着色剂
flammability	n.	易燃；可燃性
lubricant	n.	润滑剂
mica	n.	[矿]云母
pigment	n.	[生]色素，颜料
polycarbonate	n.	[化]聚碳酸酯
polymethylmethacrylate		聚甲基丙烯酸甲酯，有机玻璃
polyvinyl	adj.	[化]乙烯聚合物的
thermoplastics		热塑性塑料
weatherability	n.	耐气候性；抗风化力

（一）课文导读

本文从单体、结合键、树枝状和网状这三个方面详细介绍了塑料的结构。塑料由聚合分子和各种不同的添加剂组成。聚合物是长链分子（也称巨大分子或高分子），它是通过聚合形成的，即单体的结合和交联。

（二）课文参考译文

塑料的结构

塑料由聚合分子和各种添加剂组成。聚合物是长链分子（也称巨大分子或高分子），它是通过聚合形成的，即单体的结合和交联。

单体

单体是聚合物基本的结构单元。mer 这个词（来自希腊语的 meros，意思是"部分"）

表示最小的重复个体，与晶体结构中的术语"单位晶格"相似。因此，聚合物意味着许多链节或单元。单体是有机材料，它们包含碳原子，其中的碳原子与其他原子如氢、氧、氮、氟、氯、硅和硫以共价键（电子共享）结合在一起。

一种典型的单体是如图 14-1（a）所示的乙烯分子。这些分子可以通过加热、加压和添加催化剂的化学反应过程，把它们自己依附在其他乙烯分子上。在这个反应中，碳原子之间的双键打开，分子排列成长线形式［图 14.1（b）］。

图 14.1（b）所示的聚合物称为线型聚合物，因为其结构是线型的。分子链中分子的数量（链的长度）称为分子量。线型分子并不意味着它是直的。这些分子（链）有不同的长度，而且它们的排列是非晶态的（无任何长程有序）。这种排列经常被描述为一碗意大利式细面条，或者是水桶里的蠕虫，全都相互缠绕。因为分子链的长度不同，所以聚合物的分子量是一个统计值。

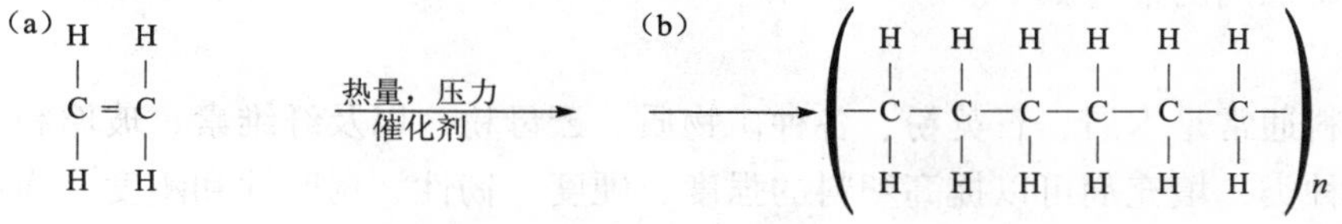

图 14-1 聚合物分子的基本结构示意图
(a) 乙烯分子；(b) 聚乙烯分子

结合键

虽然在每个长链分子之间有一个共价键（主键），但是在不同的链之间则不存在这种分子间的共价键。在相同链之间的重叠部分存在的键被称为次键。次键是范德瓦尔斯键、氢键和离子键。所有的次键都比分子中的共价键弱得多。

两种结合键在强度上的差异有一至两个数量级。这种差异很重要，因为较弱的次键决定了聚合物的总强度。（例如，一个物体的强度取决于它最弱的部分。）

如果一条分子链中的重复单元都是同一类型，那么这种高分子称为均聚物。但是，为了获得某些特殊性能，一种聚合物中可以结合两种或三种不同类型的单体，分别称为共聚物和三元共聚物。

树枝状和网状

共聚物的性能不仅取决于两个单体，也取决于它们在分子结构中的排列（图 14-2）。例如，树枝状［图 14-2（b）］会干扰分子的相对运动，从而影响变形抗力。另一种分子链形式是交联［图 14-2（c）］，它是相邻分子链联结在一起形成的空间（三维）网络结构。交联对聚合物的性能有显著影响，例如橡胶的硫化。

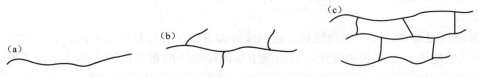

图 14-2 聚合物分子链示意图
(a) 线型结构；(b) 树枝状结构；(c) 网状结构

热塑性塑料，例如丙烯酸树脂、尼龙、聚乙烯和聚氯乙烯，是线型分子。热固性塑料，例如环氧树脂、酚醛塑料和聚硅酮，具有交联型的分子链。交联的一个例子是用于汽车轮胎的橡胶的硫化，每一个轮胎就是一个巨大分子。

具有线型分子链结构的塑料称为热塑性塑料，由交联型分子链构成的塑料称为热固性塑料。

阅读材料（7）参考译文

添 加 剂

为了获得某些特殊性能，聚合物中通常加入添加剂。添加剂可以改变或者提高某些特性，如刚度、强度、颜色、耐气候性、可燃性、针对电气应用的耐电弧性和后续加工的简易性。对这些添加剂的描述如下。

填充剂

这些材料通常是木屑、石英粉、各种矿物质、云母粉，以及纤维素、玻璃和石棉的短纤维。根据其种类，填充剂可以提高塑料的强度、硬度、韧性、耐磨性和刚度。通过不同类型聚合物/填充剂的各种组合百分比使得这些性能达到最大值。

填充剂的有效性取决于填充剂与分子链之间的结合键的本质。由于填充剂的成本较低，所以填充剂也降低了聚合物单位重量的总成本。大部分热塑性塑料和一些热固性塑料中含有添加剂。

增塑剂

增塑剂通过降低玻璃态转变温度来增加聚合物的柔顺性和柔软度。增塑剂是低分子量的溶剂（具有高沸点，即非挥发性），可以降低聚合物长链分子之间次键的强度，从而使聚合物变得柔软。最常应用于聚氯乙烯（PVC），并使PVC保持柔软性。

稳定剂

许多聚合物受到紫外线和氧气的不利影响，紫外线和氧气会削弱和破断长链分子的主键，然后聚合物会变硬和变脆，这个过程称为降解。抵御紫外线的一个典型例子是橡胶与炭黑（煤烟）的混合。炭黑吸收大部分紫外线。向聚合物中添加抗氧化剂可以抵御氧气引起的降解，特别是在高温下。各种涂层是抵御降解的另一种方法。

着色剂

添加着色剂可使塑料具有各种各样的颜色。着色剂要么是有机的（染料），要么是无机的（色素）。着色剂的选择取决于工作温度和在光线中的暴露程度。色素（分散颗粒）对温度和光线的抵抗力通常高于染料。

阻燃剂

如果温度足够高，聚合物会燃烧。根据成分如含铬量和含氟量的不同，聚合物的可燃性（支持燃烧的能力）有显著的变化。例如聚甲基丙烯酸甲酯点燃后继续燃烧，但是聚碳酸酯会自己熄灭。用可燃性较低的原料制备聚合物，或者向聚合物中加入阻燃剂可以降低聚合物的可燃性。常用的阻燃剂是氯、溴和磷的化合物。

润滑剂

聚合物中加入润滑剂是为了减小随后加工和使用过程中的摩擦，以及防止黏模。为了防止聚合物薄膜黏在一起，润滑也很重要。

聚合物共混体

为了使聚合物具有橡胶态性能，可以将聚合物与少量的橡胶态聚合物混合在一起。橡胶态聚合物细小地分散在聚合物中，提高其冲击强度。这些聚合物称为橡胶改性聚合物。

Lesson 15

Classification and Application of Plastics

Plastic are generally divided into the categories of thermoplastic and thermosets plastic.

Thermoplastics

We noted earlier that, in the amorphous structure of a polymer, the bonds between adjacent long-chain molecules (secondary bonds) are much weaker than the covalent bonds (primary bonds) within each molecule. Hence, it is the strength of the secondary bonds that determines the overall strength of the polymer. Linear and branched polymers have weak secondary bonds.

If we now raise the temperature of this polymer above the glass-transition temperature, we find that it becomes softer and easier to form or to mold into a shape. The mobility of the long molecules (thermal vibrations) increases at T_g and above. If this polymer is now cooled, it returns to its original hardness and strength. In other words, the process is reversible.

Polymers that exhibit this behavior are known as thermoplastics. Typical examples are acrylics, nylon, polyethylene, and polyvinyl chloride.

Thermosets

When the long-chain molecules in a polymer are cross-linked in a three-dimensional (spatial) network, the structure becomes in effect one giant molecule with strong covalent bonds. Cross-linking is done by the polymerization processes. Because of the nature of the bonds, the strength and hardness of such a polymer is not affected by temperature.

These polymers are known as thermosets because, during polymerization under heat and pressure, the network is completed and the shape of the part is permanently set (curing). Unlike in thermoplastics this reaction is irreversible and the polymer cannot be recycled. However, if the temperature is increased sufficiently the thermosetting polymer begins to decompose, char, and degrade. Thermosetting polymers do not have a sharply defined glass-transition temperature. Commonly, thermosetting resins become rubbery and compliant across a narrow temperature range.

The response of a thermosetting plastic to temperature can be likened to boiling an egg or baking a cake. Once the cake is baked and cooled, reheating it will not change its shape, and if the temperature is too high, it will burn. On the other hand, the response of a thermoplastic can be likened to ice cream. It can be softened, refrozen, and resoftened a number of times. It can be molded into shapes, frozen, and then softened again to be remolded into a different shape.

A typical example of a thermosetting plastic is phenolic, which is a product of the reaction between phenol and formaldehyde. Typical products of this polymer are the handles on cooking pots and pans and electrical components such as switches.

The polymerization process for thermosets generally takes place in two stages. The first one is at the chemical plant, where the molecules are partially polymerized into linear chains. The second stage is at the parts-producing plant, where the cross-linking is completed under heat and pressure during the molding of the part.

Thermosetting plastics generally possess better mechanical, thermal, chemical, and electrical resistance and better dimensional stability than thermoplastics.

Words and Expressions

acrylics [əˈkriliks]		丙烯酸树脂
bake [beik]	v.	烘焙，烤，烧硬
boiling [ˈbɔiliŋ]	adj.	沸腾的；激昂的
char [tʃɑː]	v.	烧焦
degrade [diˈgreid]	v.	(使)降级；(使)堕落；(使)退化
irreversible decompose		不可逆的分解
mobility [məuˈbiliti]	n.	活动性，灵活性；迁移率；机动性
nylon [ˈnailən]	n.	尼龙
permanentlyset [pəˈmæŋgənit]	n.	[化]高锰酸
phenol formaldehyde		苯酚甲醛
phenolic [fiˈnɔlik]	adj.	[化]酚的；石碳酸的
polyethylene [ˌpɔliˈeθiliːn]	n.	[化]聚乙烯
polymerization [ˌpɔliməraiˈzeiʃən]	n.	聚合
polyvinyl chloride	n.	聚氯乙烯
reaction [ri(ː)ˈækʃən]	n.	反应；反作用；反动(力)
reversible [riˈvəːsəbl]	adj.	可逆的
stability [stəˈbiliti]	n.	稳定性
thermal vibration		热振动

Notes

1. Plastics are generally divided into the categories of thermoplastics and thermosets plastics.
译文：塑料一般分为热塑性塑料和热固性塑料两大类。

2. We noted earlier that, in the amorphous structure of a polymer, the bonds between adjacent long-chain molecules (secondary bonds) are much weaker than the covalent bonds (primary bonds) within each molecule.

译文：我们在前面已经提到过，在聚合物的分子结构中，相邻的长链分子之间的键（次键）的结合力要弱于长链内部分子之间的共价键（主键）。

解析：句中 that 是从属连词，引导宾语从句，作动词 note 的宾语。much weaker 中 much 放在形容词的比较级之前表程度，译为"……得多"。

3. If we now raise the temperature of this polymer above the glass-transition temperature, we find that it becomes softer and easier to form or to mold into a shape.

译文：如果现在提高这种聚合体的温度至玻璃化转换温度之上，我们会发现聚合体变得更软且更容易成型或模塑成型。

解析：此句含有一个 if 引导的条件状语从句，另外主句中还包含一个由 that 引导的宾语从句，作 find 的宾语。

4. When the long-chain molecules in a polymer are cross-linked in a three-dimensional (spatial) network, the structure becomes in effect one giant molecule with strong covalent bonds.

译文：聚合体中长链分子以三维（空间）网络交联，其结构变成一个带有许多强大共价键的巨型分子。

解析：句中 in effect 的意思是"有效"。cross-link：交联。

5. Because of the nature of the bonds, the strength and hardness of such a polymer is not affected by temperature.

译文：由于结合键的本性，这种聚合物的强度和硬度不受温度影响。

解析：此句中 because of 等同于 due to，译为"由于……原因"。

6. These polymers are known as thermosets because, during polymerization under heat and pressure, the network is completed and the shape of the part is permanently set (curing). Unlike in thermoplastics this reaction is irreversible and the polymer cannot be recycled.

译文：这种聚合物被称为热固性塑料，因为在加热和加压下进行的聚合过程中，网状结构被形成，产品的形状被永久地固定下来。这种反应是不可逆的，并且这种聚合物也不能被回收，这与热塑性塑料不一样。

解析：句中 unlike、irreversible 中的 un-、ir-均为否定前缀，表示"不"，如 unable（不能的，不会的）、unacceptable（无法接受的，不受欢迎的）、irreal（不真实的，虚构的）、irregular（不规则的，无规律的）。

7. The response of a thermosetting plastic to temperature can be likened to boiling an egg or baking a cake.

译文：热固性塑料对温度的反应可比作煮蛋或烤面包。

解析：短语 be likened to 中，to 为介词，后接动词时须用动名词形式。类似的短语还有 look forward to（期待，期望，盼望）等。

8. A typical example of a thermosetting plastic is phenolic, which is a product of the reaction between phenol and formaldehyde.

译文：热固性塑料的典型实例是酚醛塑料，它是（苯）酚和甲醛之间反应的产物。

解析：句中 which 是关系代词，引导非限制性定语从句。

9. The first one is at the chemical plant, where the molecules are partially polymerized into linear chains.

Chapter 4 Plastics Forming and Mold Design

译文：第一个阶段在化工厂里进行，分子被部分聚合成线型分子链。

解析：句中 where 是关系副词，引导非限制性定语从句。

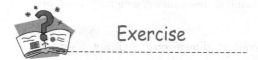

Fill in the blanks according to the text.

1. If we now raise the temperature of this polymer above the glass-transition temperature, we find _____ it becomes softer and easier to form or to mold into a shape.
2. _____ the nature of the bonds, the strength and hardness of such a polymer is not affected by temperature.
3. However, _____ the temperature is increased sufficiently the thermosetting polymer begins to decompose, char, and degrade.
4. Once the cake is baked and cooled, reheating it will not change its shape, and if the temperature is too _____, it will burn.
5. A typical example of a thermosetting plastic is phenolic, _____ is a product of the reaction between phenol and formaldehyde.

 (8)

Average Plastics

Plastic are generally divided into the categories of thermoplastic and thermosets plastic.

Thermoplastics

ABS (acrylonitrile-butadiene-styrene): Very tough, yet hard and rigid; fair chemical resistance; low water absorption, hence good dimensional stability; high abrasion resistance; easily electroplated.

Acetal: Very strong, stiff engineering plastic with exceptional dimensional stability and resistance to vibration fatigue; low coefficient of friction; high resistance to abrasion and chemicals; retains most properties when immersed in hot water.

Acrylic: High optical clarity; excellent resistance to outdoor weathering; hard, glossy surface; excellent electrical properties, fair chemical resistance; available in brilliant, transparent colors.

Cellulosics: Family of tough, hard materials, cellulose acetate, propionate, butyrate, and ethyl cellulose. Properly ranges are broad because of compounding; available with various degrees of weather, moisture, and chemical resistance; poor to fair dimensional stability; brilliant colors.

Fluoroplastics: Large family (PTFEM, FEP, PFA, ETFE, and PVF_2) of materials characterized by excellent electrical and chemical resistance, low friction, and outstanding stability at high temperatures; strength is low to moderate; cost is high.

Nylon: Outstanding toughness and wear resistance; low coefficient of friction; excellent electrical properties and chemical resistance; generally poor dimensional stability (varies among different types, however).

Phenylene Oxide: Excellent dimensional stability (very low moisture absorption) superior mechanical and electrical properties over a wide temperature range; resists most chemicals but is attacked by some hydrocarbons.

Polycarbonate: Highest impact resistance of any rigid, transparent plastic; excellent outdoor stability and resistance to creep under load; fair chemical resistance; some aromatic solvents cause stress cracking.

Polyester: Excellent dimensional stability, electrical properties, toughness, and chemical resistance, except to strong acids or bases; not suitable for outdoor use or for continuous service in water over 125 ℉ (51.7 ℃).

Polyethylene: Wide variety of grades; low-, medium- and high-density formulations. LD (low-density) types are flexible and tough. MD and HD types are stronger, harder, and more rigid; all are lightweight, easy to process, low-cost materials; poor dimensional stability and heat resistance; excellent chemical resistance and electrical properties.

Polymide: Outstanding resistance to heat - 500 ℉ (260.2 ℃) in continuous use, 900 ℉ (482.6 ℃) intermittent - and to heat aging. High impact strength and wear resistance; low Coefficient of thermal expansion; difficult to process by conventional methods; high cost.

Polyphenylene Sulfide: Outstanding chemical and heat resistance 450 ℉ (232.4 ℃) in continuous use; excellent low-temperature strength; inert to most chemicals over a wide temperature range; inherently flame-retardant; requires high processing temperature.

Polypropylene: Outstanding resistance to flex and stress cracking; excellent chemical resistance and electrical properties; good impact strength above 14° (-10 ℃); good thermal stability below 225 ℉ (107.3 ℃); light weight, low cost, can be electroplated.

Polystyrene. Low cost, easy to process, rigid, crystal-clear, brittle material; low heat resistance, poor outdoor stability; often modified to improve heat or impact resistance.

Polysulfone: Highest heat-deflection temperature of melt-processible thermoplastics; requires high processing temperature; tough, strong, and stiff; excellent electrical properties, even at high temperature; can be electroplated; high cost.

Polyurethane: Tough, extremely abrasion- and impact-resistant material; good electrical properties and chemical resistance; can be made in solid moldings or flexible foams; UV exposure produces brittleness, lower properties, and yellowing; also made in thermoset formulations.

Polyvinyl Chloride: Many formulations available, but most are classed as either rigid or flexible; rigid grades are hard, tough, and have excellent electrical properties, outdoor stability, and resistance to moisture; flexible grades (containing plasticizer) are easier to process but have lower properties; heat resistance is low to moderate for most types of PVC; low cost.

Thermosets

Alkyds: Good dimensional stability; very good dielectric properties; resistance to 300 °F (149 °C) in continuous use; high cost.

Epoxies: Excellent strength and toughness; outstanding adhesion to many materials; good resistance to many acids, alkalis and solvents; versatility and case of processing; high cost.

Melamines: Very high hardness; resistant to detergents, water, and staining; permanency of color and molded in designs; fair dimensional stability; low impact strength.

Phenolics: Resistant to 300 °F (149 °C) with some formulations up to 600 °F (315.8 °C) over an extended period; outstanding resistance to deformation over load and over wide temperature range; chemically resistant to common solvents, weak acids, and many detergents. Low cost and ease in processing; some color limitations and low impact strength.

Urethanes: Good dielectric properties; very good wear resistance and toughness; retain properties over a wide temperature range.

Words and Expressions

acetal	n.	[化]乙缩醛，乙缩醛二乙醇
alkyl	n.	[化]烷基；烃基
	adj.	烷基的；烃基的
cellulosis	adj.	纤维分解
crystal-clear	adj.	透明似水晶的；易懂的
epoxies	adj.	环氧的
fluoroplastic	n.	氟塑料
melamine	n.	[化]三聚氰胺
phenylene oxide		[化]亚苯基氧化物
polyurethane	n.	[化]聚亚氨酯
sulfone	n.	[化]砜〈美〉=sulphone
urethanes	n.	聚氨酯橡胶

课文导读 及 参考译文

（一）课文导读

本文详细介绍了塑料的分类及应用。塑料一般分为热塑性塑料和热固性塑料。其中热固性塑料一般比热塑性塑料具有更好的力学性、耐热性、化学稳定性、电绝缘性以及更好的尺寸稳定性。

（二）课文参考译文

塑料的分类与应用

塑料一般分为热塑性塑料和热固性塑料两大类。

热塑性塑料

我们在前面已经提到过，在聚合物的分子结构中，相邻的长链分子之间的键（次键）的结合力要弱于长链内部分子之间的共价键（主键）。因此，次键的强度决定了聚合物的总强度。线型和树枝状聚合物具有弱次键。

如果现在提高这种聚合物的温度至玻璃化转换温度（T_g）之上，我们会发现聚合体变得更软且更容易成型或模塑成型。在 T_g 及以上温度，长链分子的活动性（热振动）增大。如果这种聚合物现在冷却，则回到其初始硬度和强度状态。换句话说，这个软化和变硬的过程是可逆的。

具有这种性能的聚合物称为热塑性聚合物。典型的例子有丙烯酸树脂、尼龙、聚乙烯和聚氯乙烯。

热固性塑料

聚合体中长链分子以三维（空间）网络交联，其结构变成一个带有许多强大共价键的巨型分子。交联通过聚合过程完成。由于结合键的本性，这种聚合物的强度和硬度不受温度影响。

这种聚合物被称为热固性塑料，因为在加热和加压下进行的聚合过程中，网状结构被形成，产品的形状被永久地固定下来。这种反应是不可逆的，并且这种聚合物也不能被回收，这与热塑性塑料不一样。但是，如果温度升到足够高，热固性塑料会开始分解、烧焦和降解。热固性塑料没有明确定义的玻璃态转变温度。热固性树脂通常在穿过一个较窄的温度范围时变成橡胶态和柔软状。

热固性塑料对温度的反应可比作煮蛋或烤面包。一旦面包烤好和冷却，重新加热面包不会改变其形状，而且如果温度太高，面包会烧焦。另一方面，热塑性塑料对温度的反应可比作冰激凌。冰激凌可以反复多次地软化、再冷冻和再软化。热塑性塑料能模塑成型、冷却然后再次软化和再次模塑成型为不同形状。

热固性塑料的典型实例是酚醛塑料，它是（苯）酚和甲醛之间反应的产物。这种聚合物的典型制品是蒸煮罐和炊锅上的手柄，以及电气元件，如开关。

热固性塑料的聚合过程通常分两个阶段进行。第一个阶段在化工厂里进行，分子被部分聚合成线型分子链。第二阶段在制品制造厂里进行，在制品的模塑过程中，在热和压力的作用下完成交联。

与热塑性塑料相比，热固性塑料通常具有更好的力学性、耐热性、化学稳定性和电绝缘性，以及更好的尺寸稳定性。

（三）阅读材料（8）参考译文

常用塑料

塑料通常分为热塑性塑料和热固性塑料两大类。

热塑性塑料

ABS（丙烯腈-丁二烯-苯乙烯）：非常韧，还很刚硬；中等耐化学性；低吸水率，因此具有好的尺寸稳定性；高耐磨性；容易电镀。

乙缩醛：强度、刚性非常好的工程塑料，具有特别的尺寸稳定性和抗振动疲劳性；低摩擦系数；高耐磨损和耐化学药品性；当浸泡在热水中时仍保持大部分性能。

丙烯酸树脂：高透光度；优异的抗室外老化性；硬质、光滑的表面；优异的电性能，中等耐化学性；可以有鲜艳、透明的颜色。

纤维素塑料：韧、硬的一类材料，有乙酸纤维素、丙酸酯、丁酸酯和乙基纤维素。由于混合，所以适宜的范围广；可以具有不同程度的耐候性、耐湿性和耐化学性；低到中等的尺寸稳定性；鲜艳的颜色。

氟塑料：具有优异电阻和耐化学性的一大类材料（PTFEM、FEP、PFA、ETFE 和 PVF_2），低摩擦，优异的高温稳定性，低到中等强度；成本高。

尼龙：优异的韧性和耐磨性；低摩擦系数；优异的电学性能和耐化学性；尺寸稳定性通常较差（但是随类型的不同而不同）。

亚苯基氧化物：优异的尺寸稳定性（非常低的吸湿性），在较宽的温度范围内具有优良的力学和电学性能；耐大多数化学药品但是能被某些烃类侵蚀。

聚碳酸酯：在刚性、透明的塑料中其冲击抗力最高；优异的户外稳定性和载荷作用下的蠕变抗力；中等耐化学性；某些芳香烃溶剂引起应力开裂。

聚酯：优异的尺寸稳定性、电学性能、韧性，以及除强酸强碱之外的耐化学性；不适于在户外使用或在高于 125 °F（51.7 ℃）的水中连续使用。

聚乙烯：等级多；低、中和高密度类型。LD（低密度）类型柔韧。MD 和 HD 类型的强度、硬度和刚性更高；所有的聚乙烯都是轻质、容易加工、低成本的材料；低尺寸稳定性和耐热性；优异的耐化学性和电学性能。

聚酰亚胺：优异的耐热性——连续使用时 500 °F（260.2 ℃），间歇使用时 900 °F（482.6 ℃）——和耐热老化性。高冲击强度和耐磨性；低热膨胀系数；传统方法难以加工；高成本。

聚苯硫醚：连续使用时优异的耐化学性和耐热性 450 °F（232.4 ℃）；优异的低温强度；在较宽的温度范围内对大多数化学药品不敏感；固有的阻燃性；要求高加工温度。

聚丙烯：优异的抗挠裂和抗应力开裂性；优异的耐化学性和电学性能；14 °F（-10 ℃）以上良好的冲击强度；225 °F（107.3 ℃）以下良好的热稳定性；轻质、低成本、可以电镀。

聚苯乙烯：低成本，容易加工，刚性好，透明似水晶，脆性大；低耐热性，低户外稳定性；常被改性以提高耐热性或抗冲击性。

聚砜：在熔体可加工的热塑性塑料中具有最高的热变形温度；要求高加工温度；韧性、强度和刚性好；甚至在高温下也具有优异的电性能；能电镀；成本高。

聚氨酯：韧性好，极耐磨和耐冲击的材料；良好的电性能和耐化学性；能制成固体造型或软质泡沫塑料；紫外线照射会变脆，降低性能并且变黄；也制成热固性塑料形式。

聚氯乙烯：有许多种类型，但是大多数分为刚性和柔性；刚性类型的硬度高，韧性好，具有优异的电学性能，户外稳定性和耐湿性；柔性类型（含有增塑剂）更容易加工，但是性能较低；大多数类型的 PVC 具有低至中等耐热性；低成本。

热固性塑料

　　醇酸树脂：良好的尺寸稳定性；极好的介电性能；连续使用时耐 300 ℉（149 ℃）；高成本。

　　环氧树脂：优异的强度和韧性；与多种材料的粘附性优异；对多种酸、碱和溶剂具有良好的抗力；加工多样性；高成本。

　　三聚氰胺：极高的硬度；耐洗涤剂、水和染色剂；颜色和模塑形状持久；中等尺寸稳定性；低冲击强度。

　　酚醛塑料：在较长时间内耐热性达到 300 ℉（149 ℃），某些类型可达 600 ℉（315.8 ℃）；在载荷作用和较宽的温度范围内具有优异的变形抗力；对普通溶剂、弱酸和许多洗涤剂具有化学抗力；低成本，容易加工；某些颜色局限性，低冲击强度。

　　聚氨酯橡胶：良好的介电性能；极好的耐磨性和韧性；在较宽的温度范围内保持性能不变。

Lesson 16

Injection Molding

This is a commonly used and versatile process for both thermoplastics and thermosets. Injection molding is essentially similar to the hot-chamber die casting process. The pellets or granules are heated in a cylinder and the melt is then forced into a split-mold chamber either by a hydraulic plunger or by the screw system of an extruder at pressures that may exceed 20 000 psi (140 MPa). Viscosity of the melt is an important parameter because of its effect on material flow and forces.

The more recent equipment for injection molding is of the reciprocating screw type. As the pressure builds up at the die entrance due to the movement of the material, the rotating screw begins to move backward under this pressure to a predetermined distance (volume of material). The screw stops rotating and is then pushed forward hydraulically, forcing the molten plastic into the die cavity.

Although for thermoplastics the molds are relatively cool, thermosets are molded in heated molds, where polymerization and cross-linking take place. In either case, after the part is solidified (in thermoplastics), or set (cured, in thermosets), the molds are opened and the part is ejected. The molds are then closed and the process is repeated. Mold temperature is controlled by water flowing through channels in the mold block.

Injection molding is a high-rate production process with good dimensional control. The molds (generally made of tool steels or beryllium-copper) may have multiple cavities so that more than one part is made in one cycle of the machine. Because of the high cost of molds, production volume should be high for this process to be economical.

Injection molding machines, which are generally horizontal, are rated according to the capacity of the mold and the clamping force on the molds, which in most machines is a few hundred tons. Typical injection-molded parts are cups and containers, various electrical and communication components, toys, and pipe fittings.

In reaction-injection molding (RIM) a mixture of two or more reactive fluids is forced under high pressure into the mold cavity. Chemical reactions take place in the mold rapidly and the polymer solidifies. Various fibers such as glass, graphite, or boron may also be used to reinforce the materials. Structural foams are also produced by a similar method (and by extrusion) using an inert gas (usually nitrogen) and resin mixture. The product consists of a rigid cellular (closed-cell) structure with a continuous solid skin as much as 0.080 in (2 mm) in thickness. There are several methods for forming structural foams with density reductions as much as 40% from the solid structure. Because stiffness is proportional to the third power of the thickness of a part, for the same weight of material used, cellular structures are stiffer than solid plastics or metals.

Molds for RIM are generally made of polished metal, and pressures and temperatures involved are low. Major applications are automotive bumpers and fenders, thermal insulators for refrigerators and freezers, and stiffeners for structural components.

Words and Expressions

be proportional to		与……成比例
beryllium-copper		铍铜
bumper ['bʌmpə]	n.	缓冲器
cellular ['seljulə]	adj.	细胞的
clamping force		锁模力
extruder [eks'truːdə]	n.	压出机；挤压机
fender ['fendə]	n.	防卫物；挡泥板
graphite ['græfait]	n.	石墨
inert [i'nəːt]	adj.	无活动的，惰性的，迟钝的
nitrogen ['naitrədʒən]	n.	[化]氮
predetermine ['priːdi'təːmin]	v.	预定，预先确定
rated ['reitid]	adj.	定价的；额定的
reciprocating [ri'siprəkeitiŋ]	adj.	往复的，来回的，交替的；互换的；摆动的
refrigerator [ri'fridʒəreitə]	n.	电冰箱，冷藏库
solidify [sə'lidifai]	v.	(使)凝固；(使)团结；巩固
stiff [stif]	adj.	硬的；僵直的
viscosity [vis'kɔsiti]	n.	黏质；黏性

Notes

1. Injection molding is essentially similar to the hot-chamber die casting process. The pellets or granules are heated in a cylinder and the melt is then forced into a split-mold chamber either by a hydraulic plunger or by the screw system of an extruder at pressures that may exceed 20 000 psi (140 MPa).

译文：注射成型在本质上和热室压铸工艺相同。球状或粒状的塑料粒子在圆柱形的加热筒中被加热熔融，然后在由液压柱塞或螺杆传动系统产生的可能超过140兆帕的压力作用下，被强行挤入拼合的模腔中。

解析：句子中 be similar to 意思是"与……相似"。either... or...：不是……就是……，

或者……或者……。

2. As the pressure builds up at the die entrance due to the movement of the material, the rotating screw begins to move backward under this pressure to a predetermined distance (volume of material).

译文：由于塑料熔体被挤到模具入口处而导致料筒内压力上升，旋转螺杆在这种压力作用下一边旋转，一边向后移动到预定的距离（材料量）。

解析：due to 表示"由于，应归于"。句中 die entrance 的意思是"模具入口处"。又如：die equipment 译为"模具"；die cavity 的意思是"模槽，型腔"。

3. The screw stops rotating and is then pushed forward hydraulically, forcing the molten plastic into the die cavity.

译文：螺杆停止转动，然后通过液压推动前进，迫使熔融的塑料进入模具型腔。

解析：此句中 pushed forward 译为"推动……前进"。

4. The molds (generally made of tool steels or beryllium-copper) may have multiple cavities so that more than one part is made in one cycle of the machine.

译文：模具（通常由工具钢或铍铜制成）可以有多个型腔，因此在机器的一个生产周期内可以生产多个工件。

解析：此句中 so that 引导目的状语从句。此句中 cycle 译为"周期"。

5. Injection molding machines, which are generally horizontal, are rated according to the capacity of the mold and the clamping force on the molds, which in most machines is a few hundred tons.

译文：注射成型机器一般是卧式的，它是依据注射量和锁模力来进行标定，大多数注射机的锁模力有几百吨。

解析：句中 which are generally horizontal 及 which in most machines is a few hundred tons 均为非限制性定语从句。

6. In reaction-injection molding (RIM) a mixture of two or more reactive fluids is forced under high pressure into the mold cavity. Chemical reactions take place in the mold rapidly and the polymer solidifies.

译文：在反应注射成型（RIM）中，两种或更多种反应流体在高压下被挤入模具型腔中。化学反应在模具型腔中迅速发生且聚合物凝固。

7. There are several methods for forming structural foams with density reductions as much as 40% from the solid structure.

译文：形成结构泡沫有几种方法，其密度与相应的固体结构相比减少差不多40%。

解析：句中 as much as 的意思是"差不多"。

8. Because stiffness is proportional to the third power of the thickness of a part, for the same weight of material used, cellular structures are stiffer than solid plastics or metals.

译文：因为刚度与工件厚度的三次幂成比例，对于相同的材料重量而言，蜂窝结构比实体塑料或金属要更刚硬些。

解析：此句中 is proportional to 译为"与……成比例"。又如：in direct proportion（成正比）；in inverse proportion（成反比）。

Exercise

Fill in the blanks according to the text.

1. Viscosity of the melt is an important parameter because of its effect _____ material flow and forces.
2. The more recent equipment for injection molding is _____ the reciprocating screw type.
3. The molds are then closed and the process is repeated. Mold temperature is controlled by water flowing _____ channels in the mold block.
4. The molds (generally made of tool steels or beryllium-copper) may have multiple cavities _____ more than one part is made in one cycle of the machine.
5. Because stiffness is proportional to the third power of the thickness of a part, for the same weight of material used, cellular structures are stiffer _____ solid plastics or metals.

Reading Materials (9)

Classification of Plastics Mold

Plastics can be processed by a variety of methods, either in a molten state or as a solid. The basic processes are extrusion, molding, casting, and forming of sheet.

Some of these processes can be used for both thermoplastics and thermosets which are shown in Table 16-1, 16-2 and 16-3.

Table 16-1 Common shaping processes for thermoplastics

	Acrylics	ABS	Cellulose Acetate	Nylon	Polycarbonate	Polyethylene	Polypropylene	Polystyrene	Polysulfone	Polytetrafluoroethylene
Compression molding			√							
Injection molding	√	√	√	√	√	√	√	√	√	

Chapter 4 Plastics Forming and Mold Design

continued

	Acrylics	ABS	Cellulose Acetate	Nylon	Polycarbonate	Polyethylene	Polypropylene	Polystyrene	Polysulfone	Polytetrafluoroethylene
Extrusion	✓	✓	✓	✓	✓	✓	✓	✓	✓	✓
Rotational molding		✓		✓		✓	✓	✓		
Blow molding	✓	✓		✓	✓	✓	✓	✓	✓	
Thermoforming	✓	✓	✓		✓	✓	✓	✓	✓	
Casting	✓			✓						
Forging		✓		✓		✓	✓			
Foam molding		✓				✓	✓	✓		

Table 16-2 Common shaping processes for thermosets

		Alkyds	Aminos	Epoxies	Phenolics	Polyesters	Polyimide	Polyurethanes	Silicones
Nonreinforced	Compression molding	✓	✓	✓	✓	✓	✓	✓	✓
	Transfer molding	✓	✓		✓				✓
	Injection molding	✓	✓		✓				
	Rotational molding			✓					
	Casting	✓		✓	✓	✓		✓	✓
	Foam molding		✓			✓	✓	✓	✓
Fiber-reinforced	Compression molding	✓		✓	✓	✓	✓		✓
	Hand or spray-up			✓		✓	✓		
	Injection molding	✓			✓				✓
	Cold-press molding			✓		✓			
	Preform molding			✓		✓			
	Filament winding			✓		✓	✓		
	Pultrusion			✓		✓			

Source: After R. L. E. Brown. *Design and Manufacture of Plastic Parts*. © 1980 by John Wiley Sons. Inc. Reprinted by permission of John Wiley Sons. Inc.

Table 16-3 Comparative costs and production volumes for processing of plastics

		Equipment capital cost	Production rate	Tooling cost	Typical production volume, Number of parts						
					10	10^2	10^3	10^4	10^5	10^6	10^7
Nonrein-forced	Machining										
	Compression molding	High	Medium	High							
	Transfer molding	High	Medium	High							
	Injection molding	High	Fast	High							
	Rotational molding	Low	Slow	Low							
	Extrusion	Medium		Low							
	Casting	Low	Very slow	Low							
	Blow molding	Medium	Medium	Medium							
	Thermoforming	Low	Slow	Low							
	Forging	High	Slow	Medium							
	Foam molding	High	Medium	Medium							
Fiber-reinfor-ced	Compression molding	High	Medium	High							
	Hand lay-up		Slow	Low							
	Spray-up	Low	Slow	Low							
	Injection molding	High	Fast	High							
	Cold-press molding	High	Medium	Low							
	Preform molding	High	Medium	Medium							
	Filament winding	Low		Low							
	Pultrusion	Low		Low							

Source: After R. L. E. Brown. *Design and Manufacture of Plastic Parts*. © 1980 by John Wiley Sons. Inc. Reprinted by permission of John Wiley Sons.

Words and Expressions

blow molding	n.	吹塑（法）
casting	n.	铸件；铸造
compression mold	n.	平板硫化(法)；压缩模塑(法)；压模(法)(塑料)
extrusion	n.	挤出，推出；[地]喷出；突出；赶出
filament winding	n.	灯丝电源绕组

Chapter 4　Plastics Forming and Mold Design

课文导读 及 参考译文

（一）课文导读

本文主要介绍了注射成型普遍应用于热塑性塑料和热固性塑料的成型。其中螺杆式注塑机是用来进行注塑成型的设备。

（二）课文参考译文

注射成型

注射成型普遍应用于热塑性塑料和热固性塑料的成型。注射成型在本质上和热室压铸工艺相同。球状或粒状的塑料粒子在圆柱形的加热筒中被加热熔融，然后在由液压柱塞或螺杆传动系统产生的可能超过140兆帕的压力作用下，被强行挤入拼合的模腔中。熔体的黏度是一个重要参数，因为它影响材料的流动和作用力。

注射成型最现代的设备是往复螺杆式注塑机。由于塑料熔体被挤到模具入口处而导致料筒内压力上升，旋转螺杆在这种压力作用下一边旋转，一边向后移动到预定的距离（材料量）。螺杆停止转动，然后通过液压推动前进，迫使熔融的塑料进入模具型腔。

虽然对于热塑性塑料而言模具相对较冷，但是热固性塑料在热模中发生聚合与交联而成型。在塑件凝固（热塑性塑料）或硬化（固化，热固性塑料）后，模具打开，塑件脱模。模具然后闭合，重复工艺过程。模具温度通过沿模具滑块中通道流动的水来控制。

注射成型是高效率的生产工艺，具有良好的尺寸控制性。模具（通常由工具钢或铍铜制成）可以有多个型腔，因此在机器的一个生产周期内可以生产多个工件。由于模具成本高，该工艺的生产量应该大，以便经济实用。

注射成型机一般是卧式的，它是依据注射量和锁模力来进行标定，大多数注射机的锁模力有几百吨。典型的注射成型塑件有杯子和容器，各种电器元件和通信组件，玩具和管配件。

在反应注射成型（RIM）中，两种或更多种反应流体在高压下被挤入模具型腔中。化学反应在模具型腔中迅速发生且聚合物凝固。各种纤维如玻璃、石墨和硼也能用于增强材料。结构泡沫也采用类似的方法（并通过挤压）制备，即利用惰性气体（通常为氮气）和树脂的混合物。制品由刚性的蜂窝（闭孔型）结构和厚度为0.080英寸（2毫米）的连续的固体表面所构成。形成结构泡沫有几种方法，其密度与相应的固体结构相比减少差不多40%。因为刚度与工件厚度的三次幂成比例，对于相同的材料重量而言，蜂窝结构比实体塑料或金属要更刚硬些。

RIM（反应注射成型）所用模具通常由抛光的金属制成，而且选用的温度和压力低。主要应用于汽车的缓冲器和挡泥板，冰箱和冷藏库的热绝缘体，以及结构部件的加强筋。

（三）阅读材料（9）参考译文

塑料模的分类

塑料能以融熔态或者固态通过各种方法进行加工。基本工艺有挤压、模塑成型、铸造和

123

薄板成型。

其中一些工艺适用于热塑性塑料和热固性塑料,见表 16-1、表 16-2 和表 16-3。

表 16-1 热塑性塑料的常用成型工艺

	丙烯酸树脂	ABS	醋酸纤维素	尼龙	聚碳酸酯	聚乙烯	聚丙烯	聚苯乙烯	聚砜	聚四氟乙烯
压缩成型			√							
注射成型	√	√	√	√	√	√	√	√	√	
挤出成型	√	√	√	√	√	√	√	√	√	√
滚塑		√		√		√	√	√		
吹塑	√	√		√	√	√	√	√		
热成型	√	√	√		√	√	√	√		
浇铸	√			√						
锻造		√				√	√	√		
发泡成型		√				√	√	√		

表 16-2 热固性塑料的常用成型工艺

		醇酸树脂	氨基酸	环氧树脂	酚醛塑料	聚酯	聚酰亚胺	聚氨酯	聚硅酮类
未增强的	压缩成型	√	√	√	√	√	√	√	√
	传递成型	√	√		√				√
	注射成型	√	√		√				
	滚塑成型			√					
	浇铸	√							
	发泡成型		√						
纤维增强的	压缩模塑	√		√		√			√
	手糊或喷涂成型			√		√			
	注射成型	√			√				√
	冷压成型			√		√			
	模塑成型			√		√			
	长纤维卷绕法			√		√			
	拉挤成型			√		√			

表 16-3 塑料加工的成本和生产量比较

		设备的资金成本	生产率	工具成本	典型生产量，工件数量						
					10	10^2	10^3	10^4	10^5	10^6	10^7
未增强的	切削加工										
	压缩成型	高	中	高							
	传递成型	高	中	高							
	注射成型	高	快	高							
	滚塑成型	低	慢	低							
	挤出成型	中		低							
	浇铸成型	低	很慢	低							
	吹塑	中	中	中							
	热成型	低	慢	低							
	锻造成型	高	慢	中							
	发泡成型	高	中	中							
纤维增强的	压缩成型	高	中	高							
	手糊成型		慢	低							
	喷涂成型	低	慢	低							
	注射成型	高	快	高							
	冷压成型	高	中	低							
	模塑成型	高	中	中							
	长纤维卷绕法	低		低							
	拉挤成型	低		低							

Lesson 17

Injection Machine

The greatest quantities of plastic parts are made by injection molding. The process consists of feeding a plastic compound in powdered or granular form from a hopper through metering and melting stages and then injecting it into a mold. After a brief cooling period, the mold is opened and the solidified part ejected. In most cases, it is ready for immediate use.

Several methods are used to force or inject the melted plastic into the mold. The most commonly used system in the larger machines is the in-line reciprocating screw, as shown in Fig. 17-1.

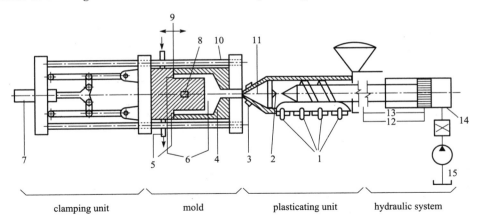

Fig. 17-1 Typical possible measuring points in the injection molding process

1. cylinder temperature; 2. melt temperature in the screw tab; 3. nozzle temperature; 4. mold wall temperature; 5. temperature of the cooling medium; 6. internal mold pressure; 7. pressure in clamping cylinder; 8. mold deformation; 9. mold breathing; 10. column elongation; 11. pressure in the screw tab; 12. screw advance speed; 13. screw displacement; 14. hydraulic pressure; 15. oil temperature

The screw acts as a combination injection and plasticizing unit. As the plastic is fed to the rotating screw, it passes through three zones as shown: feed, compression, and metering. After the feed zone, the screw-flight depth is gradually reduced, forcing the plastic to compress. The work is converted to heat by shearing the plastic, making it a semifluid mass. In the metering zone, additional heat is applied by conduction from the barrel surface. As the chamber in front of the screw becomes filled, it forces the screw back, tripping a limit switch that activates a hydraulic cylinder that forces the screw forward and injects the fluid plastic into the closed mold, An antiflowback valve prevents plastic under pressure from escaping back into the screw flights.

The clamping force that a machine is capable of exerting is part of the size designation and is measured in tons. A rule-of-thumb can be used to determine the tonnage required for a particular job. It is based on two tons of clamp force per square inch of projected area. If the flow pattern is difficult and the parts are thin, this may have to go to three or four tons.

Many reciprocating-screw machines are capable of handling thermosetting plastic materials. Previously these materials were handled by compression or transfer molding. Thermosetting materials cure or polymerize in the mold and are ejected hot in the range of 375 °F – 410 °F (190 °C – 210 °C). Thermoplastic parts must be allowed to cool in the mold in order to remove them without distortion. Thus thermosetting cycles can be faster. Of course the mold must be heated rather than chilled, as with thermoplastics.

Ways of injection molding plastic material are sketched in Fig. 17-2. The oldest is the single-stage plunger method. When the plunger is drawn back, raw material falls from the hopper into the chamber. The plunger is driven forward to force the material through the heating cylinder where it is softened and squirted under pressure into the mold. The single-stage reciprocating screw system has become more popular (Fig. 17-2) because it prepares the material more thoroughly for the mold and is generally faster. As the screw turns, it is pushed backward and crams the charge from the hopper into the heating cylinder. When enough material has been prepared, the screw stops turning and is driven forward as a plunger to ram the charge into the die.

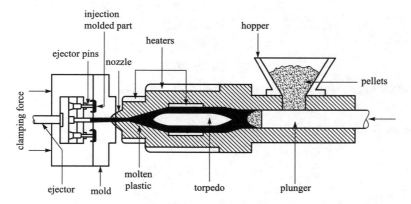

Fig. 17-2 Conventional single-stage plunger type injection molding systems

An injection molding machine heats to soften, molds, and cools to harden a thermoplastic material. Operating-temperature is generally between 150 °C and 380 °C (300 °F and 700 °F) with full pressure usually over 35 and up to 350 MPa (5 000 to 50 000 psi). The mold is water cooled. The molded piece and sprue are withdrawn from the injection side and ejected from the other side when the mold is opened. The mold is then closed and clamped to start another cycle. Thermosetting plastics can be injection molded but have to be polymerized and molded before they set in the machine. This may be done in a reciprocating screw machine where one charge at a time is brought to curing temperature. By another method, sometimes called jet molding, preforms are charged one at a time into a single-stage plunger machine.

Machines are available for molding sandwich parts. One cylinder and plunger injects a measured amount of skin material into the die, and then a second cylinder squirts the filler inside the mass. Finally, a final spurt from the first cylinder clears the core material from the sprue. The aim is to produce composites with optimum properties. Either case or core may be foamed.

Words and Expressions

barrel ['bærəl]	n.	桶
	vt.	装入桶内
be capable of		能胜任
designation [ˌdezig'neiʃən]	n.	指示，指定，选派；名称
granular ['grænjulə]	adj.	由小粒而成的，粒状的
hopper ['hɔpə]	n.	加料斗
insulator ['insjuleitə]	n.	绝缘体；绝热器
metering ['miːtəriŋ]		测量(法)；计[配]量；测定
ram [ræm]	n.	公差；撞锤；[计]随机存取内存；[计]随机存取存储器
	vi.	猛击，撞
	vt.	猛击；填塞；灌输
sandwich ['sænwidʒ, -tʃ]	n.	夹心面包，三明治
	vt.	夹入中间
semifluid [ˌsemi'fluːid]	n.	半流质
	adj.	半流质的
sprue [spruː]	n.	注入口；熔渣
squirt [skwəːt]	v.	喷出
tripping ['tripiŋ]	adj.	平稳地进行的，(说话)流畅的
	n.	绊跌；轻快的舞

Notes

1. The greatest quantities of plastic parts are made by injection molding.

译文：通过注射成型制造的塑料零件是最多的。

2. The process consists of feeding a plastic compound in powdered or granular form from a hopper through metering and melting stages and then injecting it into a mold.

译文：在这种工艺中，通过料斗加入的塑料混合粉和塑料粒子经过计量和熔化阶段，然后被注入模具中。

3. After a brief cooling period, the mold is opened and the solidified part ejected.

译文：经过短时间的冷却后，模具打开，凝固的塑件被顶出。

解析：此句中 eject 译为"顶出"。又如，eject from 译为"从……喷出"。

4. The work is converted to heat by shearing the plastic, making it a semifluid mass.

译文：这一过程中塑料通过剪切而生热，并成为粘流态。

解析：此句中 convert 译为"转换"。又如，convert into 译为"转换成"。例如：The sofa converts into a bed. 这张沙发可变成床。

5. As the chamber in front of the screw becomes filled, it forces the screw back, tripping a limit switch that activates a hydraulic cylinder that forces the screw forward and injects the fluid plastic into the closed mold. An antiflowback valve prevents plastic under pressure from escaping back into the screw flights.

译文：由于位于螺杆前面的加料室被充满，强行使螺杆后退，同时设置一个限位开关，用来激活液压油缸，迫使螺杆向前运动，将塑化的塑料注入闭合的模具。止流阀可以阻止由于螺杆回退模具中的塑料熔体产生倒流。

解析：句中 as 引导原因状语从句，主句中 tripping a limit switch that activates a hydraulic cylinder that forces the screw forward and injects the fluid plastic into the closed mold 是现在分词短语作状语，同时这个分词短语中还包含两个由 that 引导的定语从句，分别修饰先行词 switch 和 cylinder。句中 prevent from 意为："阻止，妨碍"。

6. The clamping force that a machine is capable of exerting is part of the size designation and is measured in tons.

译文：注射机的锁模力是设计产品尺寸的决定因素，用吨作为单位。

解析：句中 that 是关系代词，引导定语从句，修饰中心词 the force。

7. Many reciprocating-screw machines are capable of handling thermosetting plastic materials. Previously these materials were handled by compression or transfer molding.

译文：许多往复式螺杆注射机也能用来注射热固性塑料。以前这些材料是通过压缩成型或传递成型来处理的。

解析：句中 handle 译为"处理"。

8. The oldest is the single-stage plunger method. When the plunger is drawn back, raw material falls from the hopper into the chamber. The plunger is driven forward to force the material through the heating cylinder where it is softened and squirted under pressure into the mold.

译文：最老式的方法是单段柱塞法。当柱塞后退时，原材料从加料斗落入加料室。柱塞前行驱使材料通过加热的料筒，料筒里材料被软化并在压力下注入模具型腔。

解析：句中 where 是关系副词，引导定语从句，修饰中心词 cylinder。

9. When enough material has been prepared, the screw stops turning and is driven forward as a plunger to ram the charge into the die.

译文：当足够的塑料被预塑化之后，螺杆停止旋转，柱塞被向前驱动，将这些塑料熔体注入模具型腔中。

解析：此句中 die 是名词，意思是"硬模，冲模"。

10. Thermosetting plastics can be injection molded but have to be polymerized and molded

before they set in the machine.

译文：热固性塑料也能被注射成型，但必须在机器中聚合固化之前被注射。

解析：句中 can be molded 和 have to be polymerized 是情态动词的被动语态，其结构是"情态动词+ be+过去分词"。before 在此句中是连词，引导时间状语从句。

Fill in the blanks according to the text.

1. After a brief _____ period, the mold is opened and the solidified part ejected. In most cases, it is ready for immediate use.
2. As the plastic is fed to the rotating screw, it passes through three zones as shown: feed, _____ , and metering.
3. Thus thermosetting cycles can be faster. Of course the mold must be heated rather than chilled, as with _____ .
4. The _____ reciprocating screw system has become more popular because it prepares the material more thoroughly for the mold and is generally faster.
5. Finally, a final spurt from the first cylinder clears the core material from the sprue. The _____ is to produce composites with optimum properties. Either case or core may be foamed.

（一）课文导读

本文详细介绍了注塑机的分类、结构和特点。在比较大的机器中，最常使用的系统是直线往复式螺杆注塑机。螺杆起着混合与注入的作用。当塑料被输入到旋转螺杆时，它要经过以下三个区域：进料、压缩和计量。

（二）课文参考译文

注 塑 机

通过注射成型制造的塑料零件是最多的。在这种工艺中，通过料斗加入的塑料混合粉和塑料粒子经过计量和熔化阶段，然后被注入模具中。经过短时间的冷却后，模具打开，凝固的塑件被顶出。在大多数情况下，它可以立即使用。

有几种方法可以将融熔的塑料推入或注入模具中。在较大机器中最常使用的系统是直线往复式螺杆注射机，如图17-1所示。

螺杆起着混合与注入的作用。当塑料加入旋转的螺杆中后，它经历了三个区域：进料、压缩和计量。在进料区之后，螺杆的螺纹深度逐渐减小，迫使塑料压缩。这一过程中塑料通过剪切而生热，并成为粘流态。在计量区，通过从料筒表面传来的热量进行辅助加热。由于位于螺杆前面的加料室被充满，强行使螺杆后退，同时设置一个限位开关，用来激活液

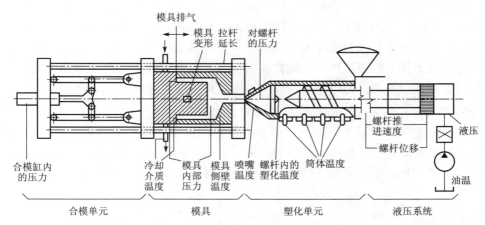

图 17-1 注射成型工艺中典型的测量点

压油缸,迫使螺杆向前运动,将塑化的塑料注入闭合的模具,止流阀可以阻止由于螺杆回退模具中的塑料熔体产生倒流。

注塑机的锁模力是设计产品尺寸的决定因素,用吨作单位。可以用经验法则来确定特定工件所需的吨位。它是以每平方英寸投影面积2吨锁模力为基础。如果塑料流动困难或者塑件薄,那么每平方英寸投影面积的锁模力可以是3~4吨。

许多往复式螺杆注射机也能用来注射热固性塑料。以前这些材料是通过压缩成型或传递成型来处理的。热固性塑料在模具中固化或聚合,在375 °F~410 °F(190 ℃~210 ℃)温度范围时脱模。为了取出时不变形,热塑性塑件必须在模具中冷却。因此,热固性塑料的生产周期更短些。当然,与热塑性塑料不同的是,其模具必须加热而不是冷却。

塑料材料注射成型的方法如图 17-2 所示。最老式的方法是单段柱塞法。当柱塞后退时,原材料从加料斗落入加料室。柱塞前行驱使材料通过加热的料筒,在料筒里的材料被软化并在压力下注入模具型腔。单段往复式螺杆系统越来越普遍(图 17-1),因为它为模具预塑化材料更细致,而且通常更快一些。当螺杆旋转时,螺杆被向后推,将塑料从加料斗推入到加热的料筒中。当足够的塑料被预塑化之后,螺杆停止旋转,柱塞被向前驱动,将这些塑料熔体注入模具型腔中。

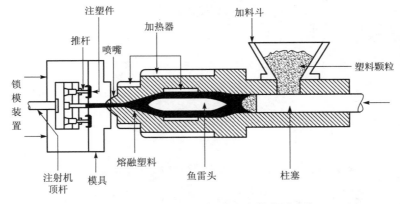

图 17-2 传统的单段柱塞式注射成型系统

注射机对热塑性塑料要加热软化、模塑成型，然后冷却硬化。操作温度通常在150 ℃ ~ 380 ℃之间（300 ℉ ~ 700 ℉），全压力通常是35 ~ 350兆帕（5 000 ~ 50 000磅/平方英尺）。模具水冷。熔体和熔渣从注射端注入，当模具打开时从另一端顶出。模具然后闭合、夹紧，开始另一个生产周期。热固性塑料也能被注射成型，但必须在机器中聚合固化之前被注射。这在往复式螺杆注射机中也可以完成，其中一次只有一个注射量达到固化温度。还有另一种方法，被称为共注射成型，按成型变化特点分层或分段注射。

有用于夹层塑件模塑成型的机器。第一个筒体和柱塞将一定量的表层材料注入模具型腔，然后第二个料筒将填充物注入材料中，最后，第一个料筒产生喷射，将心部材料从直浇道清除。该工艺的目的是为了制备性能最优的复合材料。表层和心部均可发泡。

Lesson 18

Representative Structure of Injection Mold

Molds used for injection molding of thermoplastic resins are usually flash molds, because in injection molding, as in transfer molding, no extra loading space is needed. However, there are many variations of this basic type of mold design.

The design most commonly used for all types of materials is the two-plate design (Fig. 18-1). The cavities are set in one plate, the plungers in the second plate. The sprue bushing is incorporated in that plate mounted to the stationary half of the mold. With this arrangement it is possible to use a direct center gate that leads either into a single-cavity mold or into a runner system for a multi-cavity mold. The plungers and ejector assembly and, in most cases, the runner system belong to the moving half of the mold. This is the basic design of an injection mold, though many variations have been developed to meet specific requirements.

A three-plate mold design (Fig. 18-2) features a third, movable plate which contains the cavities, thereby permitting center or offset gating into each cavity for multicavity operation. When the mold is opened, it provides two openings, one for ejection of the molded part and the other for removal of the runner and sprue.

Moldings with inserts or threads or coring that cannot be formed by the normal functioning of the press require installation of separate or loose details or cores in the mold. These loose members are ejected with the molding. They must be separated from the molding and reinstalled in the mold after every cycle. Duplicate sets are therefore used for efficient operation.

Hydraulic or pneumatic cylinders may be mounted on the mold to actuate horizontal coring members. It is possible to mold angular coring, without the need for costly loose details, by adding angular core pins engaged in sliding mold members. Several methods may be used for unscrewing internal or external threads on molded parts: For high production rates automatic unscrewing may be done at relatively low cost by the use of rack-and-gear mechanism actuated by a double-acting hydraulic long-stroke cylinder. Other methods of unscrewing involve the use of an electric gear-motor drive or friction-mold wipers actuated by double-acting cylinders. Parts with interior undercuts can be made in a mold which has provision for angular movement of the core, the movement being actuated by the ejector bar that frees the metal core from the molding.

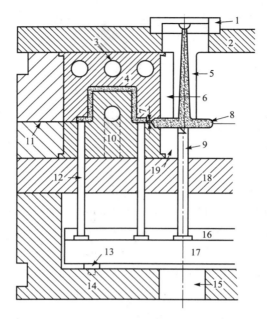

Fig. 18-1 A two-plate injection-mold design:
1. locating ring; 2. clamping plate; 3. water channels; 4. cavity; 5. sprue bushing; 6. cavity retainer; 7. gate;
8. full round runner; 9. sprue puller pin; 10. plunger; 11. parting line; 12. ejector pin; 13. stop pin;
14. ejector housing; 15. press ejector clearance; 16. pin plate; 17. ejector bar;
18. support plate; 19. plunger retainer.

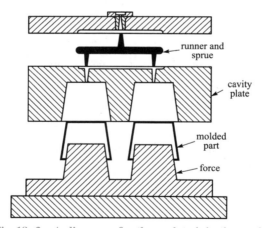

Fig. 18-2 A diagram of a three-plate injection mold

Number of Mold Cavities

Use of multiple mold cavities permits greater increase in output speeds. However, the greater complexity of the mold also increases significantly the manufacturing cost. Note that in a single-cavity mold the limiting factor is the cooling time of the molding, but with more cavities in the mold the plasticizing capacity of the machine tends to be the limiting factor. Cycle times therefore do not increase prorate with the number of cavities. There can be no clear-cut answer to the question of optimum number of mold cavities, since it depends on factors such as the complexity of the molding, the size and type of the machine, cycle time, and the number of moldings required. If a fairly accurate estimate can be made of the costs and cycle times for molds with each possible number of cavities and a cost of running the machine (with material) is assumed, a break-even quantity of the number of moldings per hour can be calculated and compared with the total production required.

Runners

The channels through which the plasticized material enters the gate areas of the mold cavities are called runners. Normally, runners are either full round or trapezoidal in cross section. Round cross section offers the least resistance to the flow of material but requires a duplicate machining operation in the mold, since both plates must be cut at the parting line. In three-plate mold designs, however, trapezoidal runners are preferred, since sliding movements are required across the parting-line runner face.

One can see from Fig. 18-2 that a three-plate mold operation necessitates removal of the runner and sprue system, which must be reground, and the material reused. It is possible, however, to eliminate the runner system completely by keeping the material in a fluid state. This mold is called a hot-runner mold (Fig. 18-3). The material is kept fluid by the hot-runner manifold, which is heated with electric cartridges. The advantage of a hot-runner mold is that in a long-running job it is the most economical way of molding-there is no regrinding, with its attendant cost of handling and loss of material, and the mold runs automatically, eliminating variations caused by operators. A hot-runner mold also presents certain difficulties: It takes considerably longer to become operational,

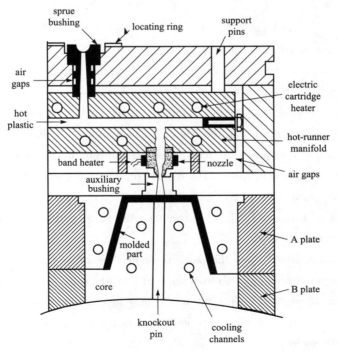

Fig. 18-3 Schematic drawing of a hot-runner mold

and in multicavity molds balancing the gate and the flow and preventing drooling are difficult. These difficulties are partially overcome in an insulated-runner mold, which is a cross between a hot-runner mold and a three-plate mold and has no runner system to regrind. An insulated-runner mold is more difficult to start and operate than a three-plate mold, but it is considerably easier than a hot-runner mold.

Gating

The gate provides the connection between the runner and the mold cavity. It must permit enough material to flow into the mold to fill out the cavity. The type of the gate and its size and location in the mold strongly affect the molding process and the quality of the molded part. There are two types of gates: large and restricted. Restricted (pinpointed) gates are usually circular in cross section and for most thermoplastics do not exceed 0.060 in. in diameter. The apparent viscosity of a thermoplastic is a function of shear rate-the viscosity decreases as the shear rate and, hence, the velocity increases. The use of the restricted gate is therefore advantageous, because the velocity of the plastic melt increases as it is forced through the small opening; in addition, some of the kinetic energy is transformed into heat, raising the local temperature of the plastic and thus further reducing its viscosity. The passage through a restricted area also results in higher mixing.

The most common type of gate is the edge gate [Fig. 18-4 (a)], where the part is gated either as a restricted or larger gate at some point on the edge. The edge gate is easy to construct and often is the only practical way of gating. It can be fanned out for large parts or when there is a special reason. Then it is called a fan gate [Fig. 18-4 (b)]. When it is required to orient the flow pattern in one direction, a flash gate [Fig. 18-4 (c)] may be used. It involves extending the fan gate over the full length of the part but keeping it very thin.

The most common gate for single-cavity molds is the sprue gate [Fig. 18-4 (d)]. It feeds directly from the nozzle of the machine into the molded part. The pressure loss is therefore a minimum. But the sprue gate has the disadvantages of the lack of a cold slug, the high stress concentration around the gate area, and the need for gate removal. A diaphragm gate [Fig. 18-4 (e)] has, in addition to the sprue, a circular area leading from the sprue to the piece. This type of gate is suitable for gating hollow tubes. The diaphragm eliminates stress concentration around the gate because the whole area is removed, but the cleaning of this gate is more difficult than for a sprue gate. Ring gates [Fig. 18-4 (f)] accomplish the same purpose as gating internally in a hollow tube, but from the outside.

When the gate leads directly into the part, there may be surface imperfection due to jetting. This may be overcome by extending a tab from the part into which the gate is cut. This procedure is called tab gating [Fig. 18-4 (g)]. The tab has to be removed as a secondary operation.

A submarine gate [Fig. 18-4 (h)] is one that goes through the steel of the cavity. It is very often used in automatic molds.

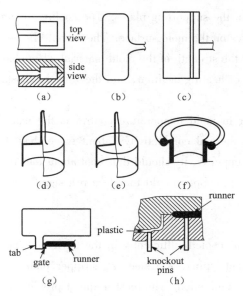

Fig. 18-4 Gating designs
(a) Edge; (b) Fan; (c) Flash; (d) Sprue; (e) Diaphragm; (f) Ring; (g) Tab; (h) Submarine

Venting

When the melted plastic fills the mold, it displaces the air. The displaced air must be removed quickly, or it may ignite the plastic and cause a characteristic burn, or it may restrict the flow of the melt into the mold cavity, resulting in incomplete filling. For venting the air from the cavity, slots can be milled, usually opposite the gate. The slots usually range from 0.001 to 0.002 in. deep and from 3/8 to 1 in. wide. Additional venting is provided by the clearance between knockout pins and their holes. Note that the gate location is directly related to the consideration of proper venting.

Parting Line

If one were inside a closed mold and looking outside, the mating junction of the mold cavities would appear as a line. It also appears as a line on the molded piece and is called the parting line. A piece may have several parting lines. The selection of the parting line in mold design is influenced by the type of mold, number of cavities, shape of the piece, tapers, method of ejection, method of fabrication, venting, wall thickness, location and type of gating, inserts, postmolding operations, and aesthetic considerations.

Cooling

The mold for thermoplastics receives the molten plastic in its cavity and cools it to solidify to the point of ejection. The mold is provided with cooling channels. The mold temperature is controlled by regulating the temperature of the cooling fluid and its rate of flow through the channels. Proper cooling or coolant circulation is essential for uniform repetitive mold cycling.

The functioning of the mold and the quality of the molded part depend largely on the location of the cooling channel. Since the rate of heat transfer is reduced drastically by the interface of two metal pieces, no matter how well they fit, cooling channels should be located in cavities and cores themselves rather than only in the supporting plates. The cooling channels should be spaced evenly to prevent uneven temperatures on the mold surface. They should be as close to the plastic surface as possible, taking into account the strength of the mold material. The channels are connected to permit a uniform flow of the cooling or heating medium, and they are thermostatically controlled to maintain a given temperature.

Another important factor in mold temperature control is the material the mold is made from. Beryllium copper has a high thermal conductivity, about twice that of steel and four times that of stainless steel. A beryllium copper cavity should thus cool about four times as fast as a stainless steel one. A mold made of beryllium copper would therefore run significantly faster than one of stainless steel.

Ejection

Once the molded part has cooled sufficiently in the cavity, it has to be ejected, This is done mechanically by KO (knockout) pins, KO sleeves, stripper plates, stripper rings or compressed air, used either singly or in combination. The most frequent problem in new molds is with ejection. Because there is no mathematical way of predicting the amount of ejection force needed, it is entirely a matter of experience.

Since ejection involves overcoming the forces of adhesion between the mold and the plastic, the area provided for the knockout is an important factor. If the area is too small, the knockout force will be concentrated, resulting in severe stresses on the part. As a result, the part may fail immediately or in later service. In materials such as ABS and high-impact polystyrene, the severe stresses can also discolor the plastic.

Sticking in a mold makes ejection difficult. Sticking is often related to the elasticity of steel and is called packing. When injection pressure is applied to the molten plastic and forces it into the mold, the steel deforms; when the pressure is relieved, the steel retracts, acting as a clamp on the plastic. Packing is often eliminated by reducing the injection pressure and/or the injection forward time. Packing is a common problem in multicavity molds and is caused by unequal filling. Thus, if a cavity seals off without filling, the material intended for that cavity is forced into other cavities, causing overfilling.

Standard Mold Bases

Standardization of mold bases for injection molding, which was unknown prior to 1940, was an important factor in the development of efficient mold making. Standard mold bases were pioneered by the D-M-E Co., Michigan, to provide the mold maker with a mold base at lower cost and with much higher quality than if the base were manufactured by the mold maker. Replacement parts, such as locating ring and sprue bushings, loader pins and bushings, KO pins and push-back pins of high quality are also available to the molder. Since these parts are common for many molds, they can be stocked by the molder in the plant and thus down time is minimized. An exploded view of the components of a standard injection-mold base assembly is shown in Fig. 18-5.

Chapter 4 Plastics Forming and Mold Design

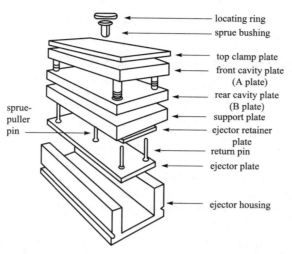

Fig. 18-5 Exploded view of a standard mold base showing component parts

Words and Expressions

diaphragm ['daiəfræm]	n.	隔板
drool [druːl]	vi.	流口水；说昏话
	vt.	从嘴淌下；散漫地说
	n.	口水；梦话
fan [fæn]	n.	扇子，风扇；迷，爱好者
	vt.	煽动；刺激；扇；吹拂
gate [geit]	n.	浇口
insulated-runner		绝热流道
jet [dʒet]	n.	喷射
	v.	喷射
KO (knockout)		推出
KO sleeve		推管
nozzle ['nɔzl]	n.	管口；喷嘴
orient ['ɔːriənt]	n.	东方
	adj.	东方的
overfilling ['əuvəˌfiliŋ]		过量灌装，超量装填
pinpoint ['pinˌpɔint]	n.	精确
	adj.	极微小的
polystyrene [ˌpɔli'staiəriːn]	n.	聚苯乙烯

regrind [ri'graind]	v.	再磨研
slug [slʌg]	n.	金属块
stripper plate		脱模板
stripper ring		推出环
submarine ['sʌbməri:n, sʌbmə'ri:n]	n.	潜水艇，潜艇
	adj.	水下的，海底的
vent [vent]	n.	通风孔；出烟孔；出口；（感情等的）发泄
	v.	放出，排出；发泄

Notes

1. This is the basic design of an injection mold, though many variations have been developed to meet specific requirements.

译文：虽然为了符合特定要求会有许多变化，但是这是注射模的基本设计。

2. Hydraulic or pneumatic cylinders may be mounted on the mold to actuate horizontal coring members.

译文：液压缸或气缸被安装在模具中以抽出侧型芯部件。

解析：此句中 be mounted on 译为"安装在……上"。

3. For high production rates automatic unscrewing may be done at relatively low cost by the use of rack-and-gear mechanism actuated by a double-acting hydraulic long-stroke cylinder.

译文：为了提高生产率，常常使用价格相对较低的长行程的液压油缸驱动齿轮齿条机构进行自动脱螺纹。

解析：句中 actuated by a double-acting hydraulic long-stroke cylinder 是过去分词短语作后置定语，修饰中心词 mechanism。

4. There can be no clear-cut answer to the question of optimum number of mold cavities, since it depends on factors such as the complexity of the molding, the size and type of the machine, cycle time, and the number of moldings required.

译文：对于模具最适宜的型腔数这个问题，没有一个确定的答案，因为这依赖于诸如塑件的复杂性、机器的类型和大小、模具的工作周期以及需要的模具的数量等方面的因素。

解析：句中 since 的意思是"因为，由于"，在此句中引导原因状语从句。depend on 的意思是"依靠，依赖"，相当于 rely on。

5. Round cross section offers the least resistance to the flow of material but requires a duplicate machining operation in the mold, since both plates must be cut at the parting line.

译文：圆形横截面对原料流动的阻力最小，但在模具上需要两次机械加工，因为在分型面上两个平板都要被切削。

解析：cross section 译为"横截面"。

6. The material is kept fluid by the hot-runner manifold, which is heated with electric cartridges.

Chapter 4　Plastics Forming and Mold Design

译文：材料通过热流道板保持液态，热流道板又是通过电子释热元件加热的。

解析：句中 which 引导非限制性定语从句。句中 electric cartridge 译为"电子释热元件"。

7. The use of the restricted gate is therefore advantageous, because the velocity of the plastic melt increases as it is forced through the small opening; in addition, some of the kinetic energy is transformed into heat, raising the local temperature of the plastic and thus further reducing its viscosity.

译文：限制性浇口的使用是有利的，因为当塑料被迫通过小浇口时，塑料熔体的速率提高，加之一些动能量转化成热量，提高了塑料的局部温度，从而进一步降低了其黏性。

8. When it is required to orient the flow pattern in one direction, a flash gate [Fig. 18-4 (c)] may be used.

译文：当要求朝一个方向流动时，可以使用薄膜浇口 [图 18-4（c）]。

解析：此句中 orient 译为"确定方向"。

9. The displaced air must be removed quickly, or it may ignite the plastic and cause a characteristic burn, or it may restrict the flow of the melt into the mold cavity, resulting in incomplete filling.

译文：型腔中原有的空气必须被迅速地排除，否则会点燃塑料并导致烧焦，或者会限定熔体进入型腔的流量而导致不完全填充。

解析：句中 displaced 是过去分词作前置定语，修饰名词 air。

10. The selection of the parting line in mold design is influenced by the type of mold, number of cavities, shape of the piece, tapers, method of ejection, method of fabrication, venting, wall thickness, location and type of gating, inserts, postmolding operations, and aesthetic considerations.

译文：影响模具设计中分型面的选择的因素有：模具的类型、型腔数目、产品的形状、拔模角度、顶出方式、模具制造工艺、排气方式、塑件的壁厚、浇口位置和类型、嵌件、后处理和外观要求。

11. Once the molded part has cooled sufficiently in the cavity, it has to be ejected, This is done mechanically by KO (knockout) pins, KO sleeves, stripper plates, stripper rings or compressed air, used either singly or in combination.

译文：一旦塑件在模具型腔中充分冷却，它就可以被推杆、推管、脱模板、推出环或压缩空气等脱模方式中的一种和多种联合作用而自动地顶出。

解析：此句中 once 是连词，意思是"一旦"。

12. Sticking is often related to the elasticity of steel and is called packing. When injection pressure is applied to the molten plastic and forces it into the mold, the steel deforms; when the pressure is relieved, the steel retracts, acting as a clamp on the plastic.

译文：黏模常常和钢的弹性有关，被称为抱紧（现象）。当注射压力作用到熔化的塑料上并迫使它注入模具的型腔时，模具钢发生弹性变形；当这种压力释放后，这种弹性变形又恢复，从而将塑料包紧。

解析：句中 relate to 的意思是"涉及，与……有关"。

13. Standardization of mold bases for injection molding, which was unknown prior to 1940, was

an important factor in the development of efficient mold making.

译文：1940年以前还无人知晓的注射成型模具模架的标准化是发展有效模具制作的重要因素。

解析：此句中 be prior to 译为"在……之前"。

Fill in the blanks according to the text.

1. The _____ provides the connection between the runner and the mold cavity.
2. Additional venting is provided by the clearance between knockout pins and their holes. Note that the gate location is directly related to the consideration of proper _____ .
3. If one were inside a closed mold and looking outside, the mating junction of the mold cavities would appear as a line. It also appears as a line on the molded piece and is called the _____ .
4. The functioning of the mold and the quality of the molded part depend largely on the location of the _____ channel.
5. Once the molded part has cooled sufficiently in the cavity, it has to be _____ .

（一）课文导读

本文从模具型腔的数量、流道、浇口、排气、分型面、冷却、推出机构和基础零件这八个方面详细介绍了注射模的典型结构。

（二）课文参考译文

注射模的典型结构

用于热塑性塑料注射成型的模具通常是溢料式模具，因为与传递模塑成型一样，在注射成型中，不需要额外的载料空间。但是，模具设计的基本类型有多种变化。

所有材料最常使用的设计是两板模设计（图18-1）。型腔装在第一个模板上，凸模装在第二个模板上。主流道衬套并入定模的模板中。按照这种安排，就有可能使用直接中心浇口，使塑料进入单腔模或者是多腔模的分流道系统中。在大多数情况下，凸模、顶出装置以及分流道系统在动模中。虽然为了符合特定要求会有许多变化，但是这是注射模的基本设计。

三板模设计（图18-2）的特点是具有包含型腔的第三个可移动的模板，因此对于多型腔操作，允许中心或偏置浇口进入每一个型腔。模具打开时有两次开模，一个是塑件的脱模，另一个是去除分流道和主流道凝料。

通过压力机常规功能不能成型的带镶嵌件、螺纹和取芯的模塑件，要求在模具中安装分离的或零散的部件或者型芯。这些零散的部件随着塑件被顶出。每一个生产周期后，这些部件必须与塑件分离，并重新安装在模具中。因此使用复制的部件以高效生产。

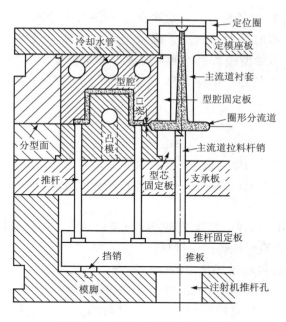

图 18-1 两板注射模设计示意图

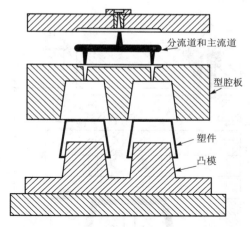

图 18-2 三板注射模示意图

液压缸或气缸被安装在模具中以抽出侧型芯部件。在模具设置料导柱，就能完成有角度侧抽芯，而不需要高成本的零散部件。有几种方法可用于旋松塑件上的内螺纹或外螺纹：为了提高生产率，常常使用价格相对较低的长行程的液压油缸驱动齿轮齿条机构进行自动脱螺纹。其他脱螺纹的方法包括使用通过双动油缸驱动的齿轮齿条机构或摩擦型弧刷。塑件的内部凸凹可以通过带料度的型芯（成型杆）成型，型芯的移动由将金属型芯与塑件分离的顶出杆驱动。

模具型腔的数目

使用多个模具型腔可以获得更高的生产效率。但是更大的模具复杂性也显著增加了制造成本。值得注意的是，单型腔模具中的限制性因素是塑件的冷却时间，但对于多型腔模具，注射机的塑化能力逐渐成为限制因素。因此，工作周期的次数不会随着型腔数目成比例增加。对于模具最适宜的型腔数这个问题，没有一个确定的答案，因为这依赖于诸如塑件的复杂性、机器的类型和大小、模具的工作周期以及需要的塑件数量等方面的因素。如果能够对具有可能型腔数目的模具的成本和工作周期次数进行相对精确的估算，并且假设机器（和材料）的运行成本，就可以计算每小时塑件数量的保本值，并与要求的总产量进行比较。

分流道

塑料熔体用以进入模具型腔的流动通道称为分流道。分流道的横截面通常是圆形或梯形的。圆形横截面对原料流动的阻力最小，但在模具上需要两次机械加工，因为在分型面上两个板都要被切削。但是，在三板模设计中，优先选用梯形分流道，因为穿过分流道面时要求有滑动。

从图18-2可以看出三板模的操作必须去除分流道和主流道冷凝料。这些凝料必须重新塑化才能重复使用。但是有可能通过保持材料为液态来完全除去分流道系统。这种模具称为热流道模（图18-3）。材料通过热流道板保持液态，热流道板又是通过电子释热元件加热的。热流道模的优点是，在长期运行工作中，它是模塑成型最经济的一种方法——没有流道凝料再次塑化，处理成本和材料损失少，而且模具自动工作，消除了人为引起的各种变化。热流道模也具有某些不足：它需要花相当长的时间才能开始运作，并且在多腔模中平衡浇口与流体，以及防止流涎是很难的。

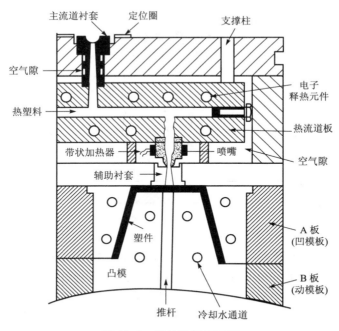

图18-3 热流道模示意图

在绝热流道模中可以部分解决这些问题，它是热流道模和三板模的结合，没有分流道系统凝料再塑化。绝热流道模比三板模要难于启动和操作，但是比热流道模要容易得多。

浇口

浇口连接着分流道与型腔。它必须允许足够的熔体流入模具来充满型腔。浇口的类型和尺寸，以及在模具中的位置都强烈地影响成型工艺和塑件的质量。浇口类型有两种：大型浇口和限制性浇口。限制性（针尖的）浇口的横截面通常是圆的，对于大多数热塑性塑料，其直径不超过 0.060 英寸。热塑性塑料的表观黏度是剪切速率的函数——黏度随着剪切速率和速度的增大而减小。限制性浇口的使用是有利的，因为当塑料被迫通过小浇口时，塑料熔体的速率提高，加之一些动能量转化成热量，提高了塑料的局部温度，从而进一步降低了其黏性。塑料通过限制性浇口时也会使混合度更高。

最常用的浇口类型是侧浇口［图18-4（a）］，通过塑件侧边上某些点处的限制性浇口或者较大浇口而进行浇铸。侧浇口在结构上容易实现，经常是唯一可行的浇口形式。对于大型塑件或者有特殊原因时侧浇口可以散开，被称之为扇形浇口［图18-4（b）］。当要求朝一个方向流动时，可以使用薄膜浇口［图18-4（c）］，它将扇形浇口延长至塑件的全长，但应保持浇口非常薄。

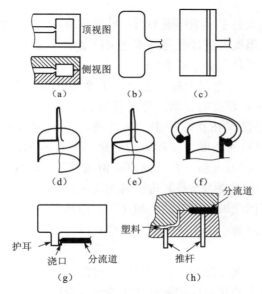

图 18-4　浇口设计示意图

(a) 侧浇口；(b) 扇形浇口；(c) 薄膜浇口；(d) 主流道浇口；(e) 隔板浇口；
(f) 环形浇口；(g) 护耳浇口；(h) 潜伏式浇口

单型腔模最常用的浇口是主流道浇口［图18-4（d）］。它从注射机喷嘴直接进料到塑件。因此压力损失最小。但是主流道浇口也存在一些不足：没有注塑冷料，浇口区域存在高的应力集中，以及需要去除浇口。隔板浇口［图18-4（e）］除了有主流道以外，还有从主流道导向塑件的圆形区域。这种浇口形式适于成型空心管。因为整个区域被移动了，所以隔板消除了浇口周围的应力集中，但这种浇口比主流道浇口要难于清理。外环形浇口［图18-4（f）］的目的与从内部成型空心管一样，只是它从外部进行浇铸。

当浇口将流体直接导入塑件时，由于喷射可能会产生表面缺陷。将护耳从塑件延长至浇口被切除处即可解决这一问题，这种工艺称为护耳浇口［图18-4（g）］。护耳必须在第二道操作中去除。

潜伏式浇口［图18-4（h）］是一种穿过型腔体的浇口，经常应用于自动模具中。

排气

当融熔的塑料充入模具时，熔体取代了空气。型腔中原有的空气必须被迅速地排除，否则会点燃塑料并导致烧焦，或者会限定熔体进入型腔的流量而导致不完全填充。为了从型腔中排出空气，通常在浇口对面开槽。沟槽通常深0.001~0.002英寸，宽3/8~1英寸。顶杆与顶杆洞之间的间隙可以进行辅助排气。值得注意的是，浇口的位置与适宜的排气直接相关。

分型面

如果从闭合模具的内部向外看，那么模具型腔的匹配面看上去像一条直线。在塑件上它看上去也像一条直线，称为分型面。一个塑件可以有几个分型面。影响模具设计中分型面的选择的因素有：模具的类型、型腔数目、产品的形状、拔模角度、顶出方式、模具制造工艺、排气方式、塑件的壁厚、浇口的位置和类型、嵌件、后处理和外观要求。

冷却

热塑性塑料的模具在型腔中容纳熔融塑料，并将其冷却凝固到顶出温度。模具具备冷却通道。通过调节冷却液的温度和流经通道的流速来控制模具温度。适当的冷却或冷却液循环对于相同的反复性的模具循环是必不可少的。

模具的功能和塑件的质量在很大程度上取决于冷却通道的位置。因为无论两个金属工件配合得多好，它们之间的界面能急剧降低热传导速率，所以冷却通道应该置于型腔中，以自己作为中心，而不是仅仅在支撑板中。冷却通道应该均匀分布以防止模具表面的温度不均匀。考虑到模具材料的强度，冷却通道应该尽可能靠近塑件表面。通道是连通的，使得冷却或加热介质可以均匀流动，并且通道是恒温控制的以保持一个特定的温度。

控制模具温度的另一个重要因素是制造模具所用的材料。铍铜的导热系数高，大约是钢的2倍、不锈钢的4倍。因此铍铜型腔的冷却速率大约是不锈钢型腔的4倍。所以由铍铜制成的模具的生产效率要远远高于由不锈钢制成的模具。

顶出机构

一旦塑件在模具型腔中充分冷却，它就可以通过推杆、推管、脱模板、推出环或压缩空气等脱模方式中的一种或多种联合作用而自动地顶出。新模具中最常见的问题就与顶出机构有关。因为没有数学的方法来预算所需要的顶出力，完全只能靠经验来确定。

因为顶出时要克服模具与塑件之间的黏附力，所以顶杆的面积是一个重要因素。如果面积太小，则顶出力集中，导致塑件上产生严重应力。结果塑件会立即或在随后的使用中损坏。在如ABS和高抗冲聚苯乙烯之类的材料中，严重的应力也会使材料褪色。

黏模会使顶出困难。黏模常常与钢的弹性有关，被称为抱紧（现象）。当注射压力作用到熔化的塑料上并迫使它注入模具的型腔时，模具钢发生弹性变形；当这种压力释放后，这种弹性变形又恢复，从而将塑料包紧。通过降低注射压力或注射时间通常可以消除抱紧现象。抱紧现象是多型腔模中的常见问题，是由不均匀填充引起的。因此，如果一个型腔没有充填就被密封，那么原本是该型腔中的材料被迫进入其他型腔，导致过量充填。

标准模架

1940年以前还无人知晓的注射成型模具模架的标准化是发展有效模具制作的重要因素。标准模架的创始人是D-M-E公司的Michigan，目的是为模具制造商提供比模具制造商自己生产的模架成本更低、质量更好的模架。置换零件如定位圈和主流道衬套、导柱和导套、高质量的推杆和拉杆等也都可以提供给模具制造商。因为这些是许多模具中的常用部件，所以可以由制造商储备在工厂里，使停工时间减到最小。标准注射模模架装配组件的分解图如图18-5所示。

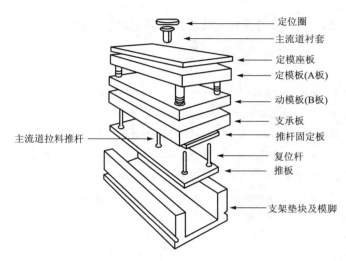

图18-5　标准模架的组件分解图

Lesson 19

Extrusion Molding

This is one of the basic methods for processing thermoplastics. The raw materials, in the form of pellets, granules, or powder, are fed into a barrel through a hopper and are extruded with a screw-type conveyor or with a plunger or ram (Fig. 19-1). Most extruders are equipped with a single screw, although two-screw (or more) extruders are also available. The materials may be heated either through internal friction during extrusion, or by external means such as a heated barrel, The process blends, compounds, homogenizes, and extrudes the plastic simultaneously at temperatures between 275 °F and 700 °F (135 ℃ to 370 ℃).

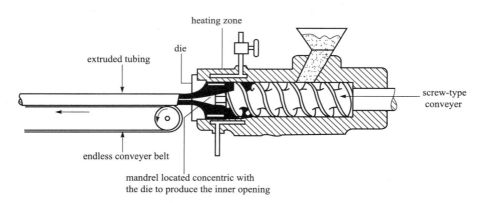

Fig. 19-1 Schematic illustration of extrusion of plastic tubing

Because this is a continuous process, long products, with various cross-sections, such as solid rods, channels, and tubing with different profiles are extruded through dies (usually tool steel) of various geometries. Sheet a few thousands of an inch in thickness is also extruded. Plastic-coated wire, cable, or strips for electrical or other applications are also extruded by this process. The wire is fed into the die opening with the extruded plastic at a controlled rate.

The extruded product is cooled by cold air, cold-water spray, or by running it through a water-filled channel. The rate and uniformity of cooling is important for dimensional control because of shrinkage and distortion. The extruded product is then coiled or cut off into desired lengths.

Pellets, which are used for other plastic-processing methods, described below, are also made by extrusion. Here the extruded product (small-diameter rod) is chopped continuously into short lengths.

Plastic-coated electrical wiring is made by a similar extrusion process. Various solid cross-sections are also extruded by this process.

 Words and Expressions

chop [tʃɔp]	n.	砍；排骨；官印；商标
	vt.	剁碎，砍；（风浪）突变
coil [kɔil]	v.	盘绕，卷
friction [ˈfrikʃən]	n.	摩擦；摩擦力
geometry [dʒiˈɔmitri]	n.	几何学
granule [ˈɡrænjuːl]	n.	小粒，颗粒，细粒
homogenize [həˈmɔdʒənaiz]	vi.	均质化
	vt.	使均匀
pellet [ˈpelit]	n.	小球
profile [ˈprəufail]	n.	剖面；侧面；外形，轮廓
shrinkage [ˈʃrinkidʒ]	n.	收缩

 Notes

1. The raw materials, in the form of pellets, granules, or powder, are fed into a barrel through a hopper and are extruded with a screw-type conveyor or with a plunger or ram.

译文：这些以粒状、球状或粉状形式存在的原材料通过加料斗被加入料筒，然后被螺旋式传送装置或柱塞或压头挤出。

2. Most extruders are equipped with a single screw, although two-screw (or more) extruders are also available.

译文：虽然也有双螺杆（或更多螺杆）挤出机，但大多数挤出机装备的是单螺杆。

3. The materials may be heated either through internal friction during extrusion, or by external means such as a heated barrel.

译文：可以通过挤出时的内摩擦，也可以通过外部途径如加热料筒来对材料进行加热。

4. Because this is a continuous process, long products, with various cross-sections, such as solid rods, channels, and tubing with different profiles are extruded through dies (usually tool steel) of various geometries.

译文：因为这是一个连续的工艺过程，诸如实心杆、水槽、各种剖面形状的管道等多种截面形状的长度较长的产品可以通过截面几何形状各异的模具（通常是工具钢）挤出。

5. Here the extruded product (small-diameter rod) is chopped continuously into short lengths.

译文：挤出的产品（小直径的细杆）被连续切成短棒。

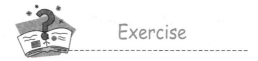 Exercise

Fill in the blanks according to the text.

1. Most extruders are equipped _____ a single screw, although two-screw (or more) extruders are also available.
2. The materials may be heated either through internal friction during extrusion, _____ by external means such as a heated barrel.
3. Because this is a continuous process, long products, with various cross-sections, _____ solid rods, channels, and tubing with different profiles are extruded through dies (usually tool steel) of various geometries.
4. The wire is fed into the die opening with the extruded plastic _____ a controlled rate.
5. The rate and uniformity of cooling is important for dimensional control _____ shrinkage and distortion.

 (10)

Section A Compression Molding

In this process, a premeasured volume of powder, or a preformed part, is placed directly in a heated mold cavity and is formed by pressure with a punch or the other half of the mold (Fig. 19-2). Compression molding is thus similar to a forging operation with the same problem of flash formation (depending on the clearances) and the need for its removal.

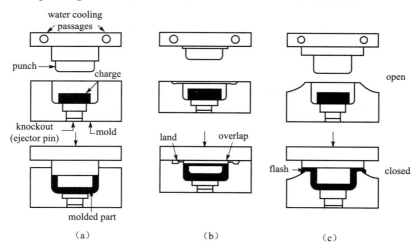

Fig 19-2 Types of compression molding
(a) flash; (b) positive; (c) semipositive

This process is used mainly for thermosets, with the original material in a partially polymerized state. Cross-linking is then completed in the heated mold under pressure and with curing times ranging from 0.5 to 5 minutes, depending on the material, part geometry and its thickness — the greater the thickness, the longer the curing time.

Because of their relative simplicity, mold costs in compression molding are generally lower than in injection molding. Three types of compression molds are available: Flash-type (for shallow or flat parts), positive (for high density), and semi-positive (for quality production), as shown in Fig. 19-2. Typical parts made by this process are knobs, handles, fittings, and housings. Fiber-reinforced materials may also be formed by compression molding.

Section B Transfer Molding

This process is a further development of compression molding. The preheated thermosetting material is placed in a transfer pot (Fig. 19-3), which is connected to the heated (about 325 °F [160 °C]) mold cavity through channels. Depending on the type of machine, a ram, plunger, or a screw feeder of an extruder forces the material (at pressures up to 12 000 psi [80 MPa]) to flow through the narrow channels into the mold cavity. This flow generates considerable heat, which raises the temperature and homogenizes the material. After filling the heated mold cavity, curing takes place by cross-linking.

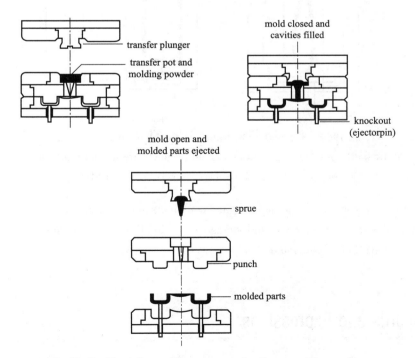

Fig 19-3 Transfer molding process for thermosetting resins.
This process is particularly suitable for intricate shapes with varying wall thickness

Transfer molding is particularly suitable for parts with intricate shapes and varying wall thicknesses and has good dimensional control. Molds are usually made of tool steel. Typical parts are electrical and electronic components.

Section C Blow Molding

This is basically a bulging process. A tubular piece of plastic is heated and then pressurized internally and expanded into the cavity of a relatively cool split mold. Typical products are hollow, thin-walled containers or articles.

An example of an automated process is extrusion blow molding. Here a hollow tube of molten plastic (parison) is produced by an extruder (more recently, by an injection-molding machine) and is then pinched at one end.

The parison is then placed into the open mold (Fig. 19-4), the mold is closed, and the parison is expanded (inflated) with an air nozzle (blowing pin) to a diameter that is usually three times the parison diameter. The air pressure is usually between 50 and 100 psi (0.35 to 0.7 MPa). The part cools off after contact with the die, which is then opened and the part is ejected. Molds are usually made of aluminum and are vented to prevent entrapment of air.

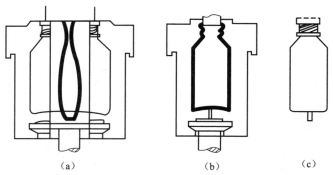

Fig. 19-4 Schematic illustration of the blow molding process.
Plastic bottles for beverage and household products are made by this process.
(a) mold open; (b) mold closed and bottle blown; (c) finished bottle

Typical products made by blow molding are containers for fluid and solid food, medicine, or other products, laboratory ware, toys, and storage tanks for liquids. Containers as large as 100 ft^3 (3 m^3) capacity have been blow molded.

Words and Expressions

hollow	n.	洞,窟窿;山谷
	adj.	空的;虚伪的;空腹的;凹的

	vi.	形成空洞
	vt.	挖空；弄凹
parison	n.	玻璃半成品；料泡；雏形
eject	vt.	逐出，摔出，驱逐；喷射
	n.	推断的事物
ware	n.	陶器，器皿

（一）课文导读

挤出成型是加工热塑性塑料的基本方法之一，是一个连续的工艺过程。诸如实心杆、水槽及各种剖面形状的管道等多种截面形状的长度较长的制品，可以通过截面几何形状各异的模具（通常是工具钢）挤出成型。

（二）课文参考译文

挤出成型

挤出成型是加工热塑性塑料的基本方法之一。这些以粒状、球状或粉状形式存在的原材料通过加料斗被加入料筒，然后被螺旋式传送装置或柱塞亦压头挤出（图 19-1）。虽然也有双螺杆（或更多螺杆）挤出机，但大多数挤出机装备的是单螺杆。可以通过挤出时的内摩擦，也可以通过外部途径如加热料筒来对材料进行加热。该工艺在 275 ℉ ~ 700 ℉（135 ℃ ~ 370 ℃）的温度范围内同时对塑料进行混合、复合、均质化并挤出塑料。

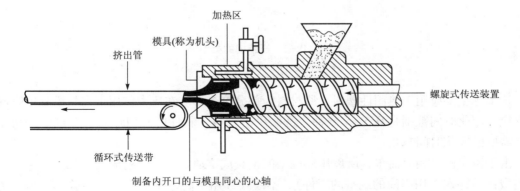

图 19-1　塑料管挤出成型示意图

因为这是一个连续的工艺过程，诸如实心杆、水槽、各种剖面形状的管道等多种截面形状的长度较长的产品可以通过截面几何形状各异的模具（通常是工具钢）挤出。厚度为几千分之一英寸的薄板也能被挤出。应用于电气和其他场合的包裹了塑料的电线、电缆和条带也是通过这种工艺挤出的。金属丝以特定的速率进入带有挤出塑料的模具中。

挤出的产品通过冷空气、冷水喷雾或者水槽来冷却。因为收缩和变形，所以冷却的速率和均匀性对于尺寸控制很重要。然后挤出的产品被盘卷或截断成所期望的长度。

用于其他塑料加工方法的塑料颗粒，也是通过挤出生产的。挤出的产品（小直径的细杆）被连续切成短棒。

塑料包裹的电气配线通过类似的挤出工艺制成。各种形状的实心横截面工件也通过该工艺挤出。

（三）阅读材料（10）参考译文

A 压缩成型

在该工艺中，将预先测量好体积的粉末或者预成型的坯件直接放入加热的模具型腔中，通过冲压机或模具另一部分的压力而成型（图19-2）。因此，压缩成型类似于锻造成型工艺，具有形成飞边（取决于间隙）并需要去除飞边的相同问题。

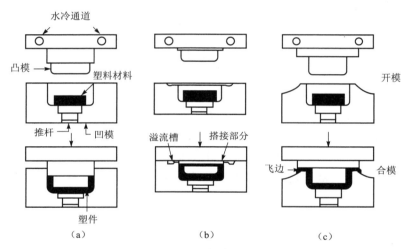

图19-2 压缩模的类型
（a）溢料式 （b）不溢式 （c）半溢式

该工艺主要应用于热固性塑料，原材料为部分聚合态。在压力作用下在热模具中完成交联反应，固化时间范围是0.5~5分钟，这取决于材料、制件的几何形状和厚度——制件越厚，需要的固化时间越长。

由于压缩模具相对简单，所以压缩成型的模具成本通常低于注射成型模具。压缩模有三种类型：溢料式（用于浅的或平的工件）、不溢式（用于高密度工件）和半溢式（用于优质化生产），如图19-2所示。压缩成型制备的典型制件有球形把手、手柄、管件和箱体。纤维增强的材料也能由压缩成型制得。

B 传递模塑

传递模塑是压缩成型的进一步发展。将预先加热的热固性塑料放入加料室中（图19-3），加料室通过通道与加热的（大约 325 ℉［160 ℃］）模具型腔相连。根据机器类型的不同，挤出机的压头、柱塞或螺杆式输送装置迫使材料（压力达 12 000 磅/平方英尺［80 兆帕］）流经窄缝通道进入模具型腔。流动过程中产生大量的热，提高了温度并使材料均质化。塑料填充完热模具型腔后，通过交联反应实现固化。

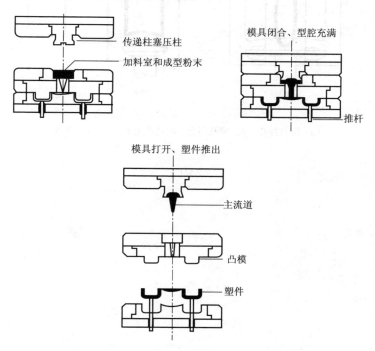

图 19-3　热固性树脂的传递模塑工艺

传递模塑工艺特别适合于形状复杂而且壁厚变化的制件，该工艺具有良好的尺寸控制性。模具通常用工具钢制成。典型的制件有电气和电子元件等。

C　吹塑成型

吹塑成型本质上是一种胀形工艺。塑料管坯首先被加热，然后内部受压，塑料膨胀后进入相对较冷的对开模具中。典型产品是空心的薄壁容器或制品。

一个自动化加工的例子是挤出吹塑成型。利用挤出机（更新一些的用注射成型机）将熔融塑料制成空心管（型坯），然后在一端箍缩。

然后把型坯放入打开的模具中（图 19-4），模具闭合，型坯通过空气喷嘴（吹杆）膨胀（胀大），膨胀后的直径通常是型坯直径的 3 倍。气压通常在 50～100 psi（0.35～0.7 兆帕）之间。制件与模具接触后冷却，然后模具打开，制件被推出。模具通常用铝合金制成，并要排气以防止夹入空气。

吹塑成型工艺生产的典型产品有用来盛装液体或固体食物、药品或其他物品的容器、实验室器皿、玩具和盛装液体的储罐。吹塑容器的容积可达 100 立方英尺（3 立方米）。

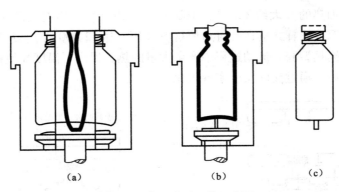

图19-4 吹塑成型工艺示意图
(a) 模具打开 (b) 模具闭合、吹制成瓶子 (c) 瓶子成品

Chapter 5

Forging Processes and Die Design
（锻造工艺及模具）

Lesson 20

Forging Processes and Die Design

Forging is one of the most important manufacturing operations. It is a plastic deformation process similar to extrusion but, unlike extrusion, it can be used to manufacture complex 3-D parts. Forging can be classified into three main categories: open-die forging, impression-die forging and closed-die forging.

Open-Die Forging

The process is schematically illustrated in Fig. 20-1. At least one of the workpiece surfaces deforms freely, and hence the open-die forging process produces parts of lesser accuracy and dimensional tolerance than impression-die or closed-die forging. However, the tooling is simple, relatively inexpensive and can be designed and manufactured with ease.

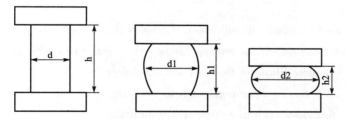

Fig. 20-1　Illustrating open-die forging

Impression-Die Forging

This forging process is schematically illustrated in Figure 20-2. It can be used to produce complex 3-D shapes having a greater accuracy than closed-die forging. The specially manufactured dies contain the negative of the forging to be produced. In one form of the process, the shape is obtained by filling the die cavity formed by the upper and lower dies. Excess material is allowed to escape into the flash. There may or may not be special provision for flash formation.

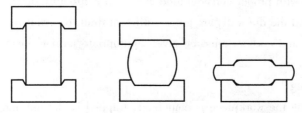

Fig. 20-2　Illustrating impression-die forging

These dies are illustrated in Fig. 20-3 and Fig. 20-4.

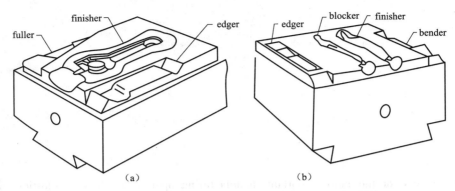

Fig. 20-3   Dies for closed-die forging (Courtesy of ASM)

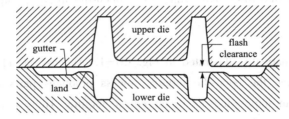

Fig. 20-4   Section through a forging-die finisher impression (Courtesy of ASM)

The die cavity must be filled without defects. Complex shape cannot be filled completely without defects, one operation starting with a rectangular or cylindrical shape. Some preforging steps are necessary to ensure complete filling of the dies without defect formation. The preforging (preforms can be produced in other forging operations such as open-die or in the same die with different die cavities, or be other processes such as rolling or roll-forging.

The preform may be further shaped to bring it closer to the final configuration in a "blocker die", which ensure proper distribution of material within the die. Excess material is allowed to run out between the dies into a flash. Before forging in the finishing die the excess material may be removed in a trimming die. A thick flash in the finishing die means high pressure within the die, which assures proper die filling. Excessive pressure may break dies or reduce their life, and some control on this may be exercised by proper control of the flash land.

The geometry of the preform and the forging dies must promote smooth material flow. Therefore, a parting line is chosen with proper consideration of the fibre structure of the finished forming. After the parting line is located the die walls are given sufficient draft to permit removal of the forging from the die cavities. Fillets and corners must be given appropriate radii to ensure smooth material flow and long die life.

Closed-Die Forging

In closed-die forging the workpiece is completely trapped in the die and no flash is generated. Material utilization is very high, but the volume of the workpiece before and after forging is identical

Chapter 5 Forging Processes and Die Design

and hence control of incoming material volume becomes critical. Excess material can create large pressures, which are liable to cause die failure.

Words and Expressions

fillet [ˈfilit]	n.	头带；带子；肉片
	vt.	用带缚；把（鱼，肉）切成片
forge [fɔːdʒ]	v.	锻造；伪造
negative [ˈnegətiv]	n.	否定；负数
	adj.	否定的；消极的；负的；阴性的
	vt.	否定
schematic [skiˈmætik]	adj.	示意性的
trap [træp]	n.	圈套，陷阱，诡计
	vi.	设圈套，设陷阱
	vt.	诱捕，诱骗，设陷阱；使受限制

1. It is a plastic deformation process similar to extrusion but, unlike extrusion, it can be used to manufacture complex 3-D parts.

译文：塑性成形工艺与挤出成形相似，但是，与挤出工艺不同的是，它能用于成形各种复杂的三维工件。

解析：此句是由 but 连接的表转折的并列句。

2. However, the tooling is simple, relatively inexpensive and can be designed and manufactured with ease.

译文：但加工很简单，费用相对便宜，而且易于设计和生产。

解析：with ease：没有困难，容易。

3. It can be used to produce complex 3-D shapes having a greater accuracy than closed-die forging.

译文：它可以用来制作比闭式模锻更为精确的复杂的三维零件。

解析：be used to：被用作……。having a greater accuracy than closed-die forging 为现在分词短语作定语，修饰中心词 shapes。

4. The preform may be further shaped to bring it closer to the final configuration in a "blocker die," which ensure proper distribution of material within the die.

译文：可以进一步把初锻坯成形，使之更接近于确保材料在模具中恰当分布的"预锻模膛"中的最后形状。

解析：句中的 which ensure proper distribution of material within the die 为定语从句，修饰

中心词 blocker die。

5. Material utilization is very high, but the volume of the workpiece before and after forging is identical and hence control of incoming material volume becomes critical.

译文：尽管原材料的使用率相当高，但由于工件锻造前后的体积是相同的，所以控制锻造坯的体积至关重要。

解析：此句是一个复合句，but 在句中起转折作用，连接两个并列句，hence 的意思为"所以"。

6. Excess material can create large pressures, which are liable to cause die failure.

译文：多余的材料会产生相当大的压力，这可能会造成模具破损。

解析：be liable to：可能干起某事来。

Fill in the blanks with the proper words.

1. In one form _____ the process, the shape is obtained _____ filling the die cavity formed _____ the upper and lower dies.
2. Some preforging steps are _____ to ensure complete _____ of the dies without _____ formation.
3. _____ pressure may break dies or reduce their life, and some control _____ this may be _____ by proper control of the flash land.
4. After the parting line is located the die walls are given _____ draft to permit removal _____ the forging _____ the die cavities.
5. Excess material can create large pressures, which are _____ cause die failure.

（一）课文导读

锻造是最重要的生产工艺之一。锻造可分为自由锻、模锻和闭式模锻，本章详细论述了它们各自的特点和原理。

（二）课文参考译文

锻造工艺及模具设计

锻造是最重要的生产工艺之一。塑性成形工艺与挤出成形相似，但是，与挤出工艺不同的是，它能用于成形各种复杂的三维工件。锻造主要有三种类型：自由锻、模锻和闭式模锻。

自由锻

该工艺如图 20-1 所示。自由锻至少有一个面可以发生自由变形，因此自由锻制造的零件在形状和尺寸精度上与模锻和闭式模锻有一定的差距。但加工很简单，费用相对便宜，而且易于设计和生产。

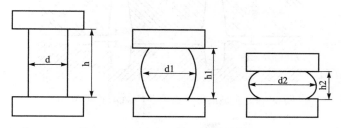

图 20-1　自由锻造过程图解

模锻

模锻过程如图 20-2 所示。它可以用来制造比闭式模锻更为精确的复杂的三维零件。这种特别加工的模具包括要加工锻件的阴模。在模锻工艺的其中一种方式中，通过填充上下模具合模形成的模腔而得到工件的形状。多余的材料可以形成飞边。对飞边的形成可以设置或不设置特殊的阻力(槽)。

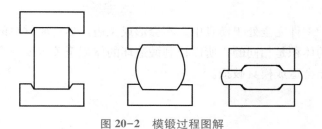

图 20-2　模锻过程图解

这些模具见图 20-3 和图 20-4。

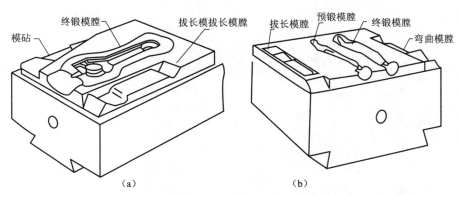

图 20-3　模锻模具图示（Courtesy of ASM）

模槽在填充的时候必须毫无缺陷。用矩形或圆柱形做坯一次锻造复杂的模腔在填充的时候不可能没有任何缺陷。因此，必须采取一些预锻步骤以确保完全填充，预防缺陷形成。预锻（预锻坯）可以在其他锻造工序进行，比如在自由锻或带不同模腔的自由锻以及轧制或辊锻过程中生产。

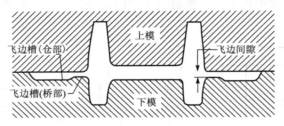

图 20-4　模锻成形切面图（Courtesy of ASM）

  可以进一步把初锻坯成形，使之更接近于确保材料在模具中恰当分布在"预锻模镗"中的最后形状。多余的材料可以向飞边槽流动形成飞边。在终锻模锻前，多余的材料可以在切边模中除去。终锻模的厚飞边意味着模内的压力很大，确保适当的模具填充。过大的压力可能会使模具断裂甚至会缩短模具的寿命，因而，必须适当地控制飞边以防止压力过大。

  预锻坯的形状和锻模必须有利于材料的顺畅流动。因此，在适当考虑成形模的物理结构的情况下要选择一个分模线。分模线确定好以后，模壁应给予足够的拔模斜度以使锻件从模腔中脱去。同时，倒角与拐角半径应足够大以保证材料的顺畅流动以及模具寿命。

闭式模锻

  在闭式模锻中，工件完全处于模具中，不会形成飞边。尽管原材料的使用率相当高，但由于工件锻造前后的体积是相同的，所以控制锻造坯的体积至关重要。多余的材料会产生相当大的压力，这可能会造成模具破损。

Chapter 6

Basic and Special Machining
（普通机械加工与特种加工）

Lesson 21

Basic Machine Tool Elements

Most machine tools are constructed by using two or more components. These components, although they may have different functions in such machines as a lathe, mill, or drill press, have common characteristics.

Because of the demand for metal removal machines such as lathes, machining centers, milling machines, grinders, and the many others shown throughout this book, there has been continuous development in flexible machining centers. The mass-produced and the special machine tools are constructed of basic elements. This chapter describes those elements.

Important requirements for machine tool structures include rigidity, shape, operator and part accessibility, ease of chip removal, and safety. In terms of machine tool performance, static and dynamic stiffness is necessary for accuracy and precision. Stability of the machine structure is required to prevent machining chatter. Understanding the basic machine elements is necessary to appreciate the breadth of modern machining methods in the manufacture of products.

Structures for Cutting Machines

Castings, forgings, and hot- or cold-formed shapes usually require machining. The variety of sizes, shapes, and materials calls for diversity in machining.

Machine tools differ not only in the number of cutting edges they employ, but also in the way the tool and workpiece are moved in relation to each other. In some machines (vertical machining centers, drill presses, boring machines, milling machines, shapers, and grinders) the workpiece remains virtually motionless and the tool moves. In others (planers, lathes, and boring mills) the tool is virtually fixed and the workpiece moves. But it should be pointed out that seldom are these simple elements applied without modification. Study Fig. 21-1 for the traditional processes used for machining parts.

The single-point tool-shaping machines are the easiest to visualize. The lathe and the boring machine are kinematic inversions employing the single-point tool. In Fig. 21-2A the work rotates in the lathe, but the cutting tool is stationary. In the boring machine Fig. 21-2B the tool rotates while the work is stationary. Although the lathe tool and the boring machine worktable are not truly stationary, this is overlooked for the moment. To feed a tool carriage past rotating work is usually more acceptable than to feed rotating work with headstock and supports past a stationary tool post.

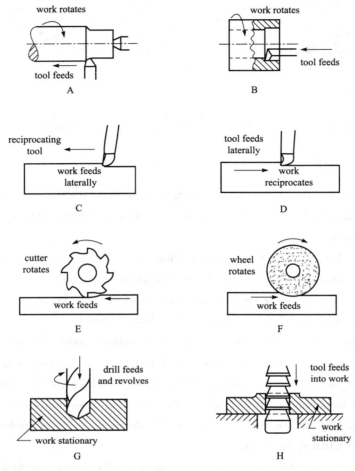

Fig. 21-1 Traditional processes used for machining
parts to specified dimensions

A. turning   B. boring   C. shaping   D. planing
E. milling   F. grinding   G. drilling   H. broaching

The shaper and planer use single-point cutting tools. Fig. 21-2C and 21-2D point out that size of the workpiece is a factor in machine structural design. The smaller workpiece is more efficiently machined on the shaper than on the planer. The general appearance of the machine is changed by reversing the kinematic relationship of work and tool. However, the cutting action principle is identical.

Before the introduction of the milling cutter by Eli Whitney in the early 1800's, the rotating tool was used only as a boring tool. But Whitney gave it a new application. The milling cutter was no longer used for cutting circular bores exclusively, but was used for cutting keyways, slitting, sawing, slab and face milling, gear cutting, and shaping irregularly formed pieces. Use of the rotating tool combined with traversing work was introduced in the milling machine as shown in Fig. 21-2E.

Chapter 6 Basic and Special Machining

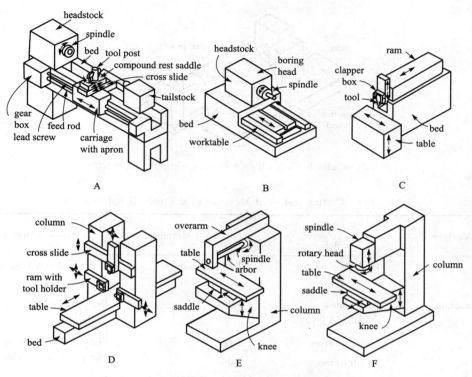

Fig. 21-2 Basic structural elements in conventional machine tools
A. lathe B. horizontal boring machine C. shaper
D. planer E. horizontal milling machine F. vertical milling

The kinematic inversion of the standard milling machine is the floor-type horizontal boring, drilling, and milling machine illustrated in Fig. 21-3 machine.

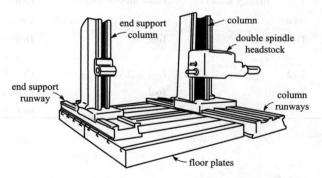

Fig. 21-3 Basic elements in a floor-type, horizontal milling, drilling, and boring machine

The cylindrical grinder adopts motions of the lathe and boring machine except for the substitution of rotating tools (the grinding wheel) for single-point tools. The work and cutting tool rotate in the grinder as shown in Fig. 21-4.

The characteristics of these basic cutting machines are listed in Table 21-1.

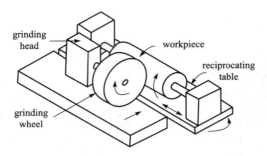

Fig. 21-4 Basic elements in a bench-type grinding machine

Table 21-1 Cutting and Feed Movement for Conventional Machines

Machine	Cutting Movement	Feed Movement	Types of Operation
Lathe	Workpiece rotates	Tool and carriage	Cylindrical surfaces, drilling, boring, reaming, and facing
Boring machine	Tool rotates	Table	Drilling, boring, reaming, and facing
Planer	Table traverses	Tool	Flat surfaces (planing)
Shaper	Tool traverses	Table	Flat surfaces (shaping)
Horizontal milling machine	Tool rotates	Table	Flat surfaces, gears, cams, drilling, boring, reaming, and facing
Horizontal boring machine	Tool rotates	Tool traverses	Flat surfaces
Cylindrical grinder	Tool (grinding wheel) rotates	Table and/or tool	Cylindrical surfaces (grinding)
Drill press	Tool rotates	Tool	Drilling, boring, facing, and threading
Saw	Tool	Tool and/or workpiece	Cut off
Broaching machine	Tool	Tool	External and internal surfaces

Words and Expressions

breadth [bredθ]　　　　　　　　　　　n.　宽度；(布的)幅宽；(船)幅
diversity [daiˈvəːsiti]　　　　　　　　n.　差异，多样性
dynamic [daiˈnæmik]　　　　　　　　adj.　动力的；动力学的；动态的
grinder [ˈgraində]　　　　　　　　　n.　磨工，尤指磨刀具的工人，研磨者

Chapter 6 Basic and Special Machining

inversion [inˈvəːʃən]	n.	倒置
kinematic [ˌkainiˈmætik]	adj.	[物]运动学的，运动学上的
machining center		加工中心
mass [mæs]	n.	块；大多数；质量；群众；大量
	adj.	群众的；大规模的；集中的
	vt.	使集合
	vi.	聚集
mill [mil]	n.	压榨机
	vt.	铣，碾磨，磨细；搅拌；使乱转
reversing [riˈvəːsiŋ]	adj.	回动的
saw [sɔː]	n.	锯
	v.	锯
slab [slæb]	n.	厚平板；厚片；混凝土路面；板层
	v.	把……分成厚片

Notes

1. Most machine tools are constructed by using two or more components.

译文：大多数机床由两个或多个部件构成。

解析：component：零件，部件。

2. In terms of machine tool performance, static and dynamic stiffness is necessary for accuracy and precision.

译文：对于机床的性能，静态和动态刚度对其精确度和精密度的影响很大。

解析：in terms of：从某方面/某角度看。

3. The variety of sizes, shapes, and materials calls for diversity in machining.

译文：由于尺寸、形状和材料不同，因此切屑加工方式也要求多样化。

解析：call for：要求，需要。

4. Machine tools differ not only in the number of cutting edges they employ, but also in the way the tool and workpiece are moved in relation to each other.

译文：机床不仅在拥有的刃具的数量上不同，而且在工具和工件的相对运动方式上也有所不同。

解析：differ in sth：在……方面不同。

5. But it should be pointed out that seldom are these simple elements applied without modification.

译文：但应该指出的是，这些简单部件很少有不经过修改而使用的。

解析：point out：指出，指明。

6. The milling cutter no longer used for cutting circular bores exclusively, but was used for cutting keyways, slitting, sawing, slab and face milling, gear cutting, and shaping irregularly-formed pieces.

译文：铣刀不再是仅仅用于加工圆孔，而且还用来加工键槽、开缝、锯削、平面铣和端铣、切齿以及成形不规则工件。

解析：no longer：不再。

7. The cylindrical grinder adopts motions of the lathe and boring machine except for the substitution of rotating tools (the grinding wheel) for single-point tools.

译文：除了采用旋转工具（砂轮）代替单刃刀具外，外圆磨床和车床、镗床的运动方式是一样的。

解析：except for：除……以外。

Fill in the blanks with the proper words.

1. With the _____ of the milling cutter by Eli Whitney in the early 1800's, the rotating tool _____ _____ only as a boring tool.

2. But Whitney gave it a new _____. The milling cutter was no longer used for cutting circular bores _____, but was used for cutting keyways, slitting, sawing, slab and face milling, gear cutting, and shaping irregularly formed pieces.

3. Use of the rotating tool _____ with traversing work was introduced in the milling machine as shown in Fig. 21-2E.

 (11)

Turning

Turning operations are most commonly associated with a machine tool known as a lathe. Evidence of the basic beginning of a lathe has been documented in Egypt reaching as far back in history as the third century B.C. Perhaps the single most important improvement in recent years was the principle of screw cutting, which was developed in France in 1740. Steady improvements to this highly versatile machine tool have occurred ever since.

Turning may be defined as a machining process by which cylindrical, conical, or irregularly shaped external or internal surfaces are produced on a rotating workpiece. The cutting action is generated by one or more stationary single-pointed cutting tools which are held at an angle to the axis of the workpiece rotation.

Description of the Process

Fig. 21-5 illustrates how the shape of the desired workpiece surface is determined by the shape and size of the stationary cutting tool, which is being fed inwardly but without longitudinal feed.

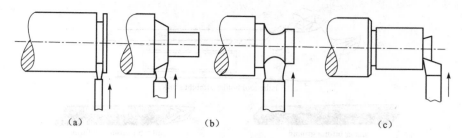

Fig. 21-5 Typical workpiece surfaces produced by in-feeding on a lathe

Speed and Feed

Speed and feed are important cutting factors. Speed is a term that describes the cutting speed or velocity of the rotating workpiece as it moves past the cutting edge of the tool. The measurement of velocity is in units of surface feet per minute or sfpm. Spindle speed, on the other hand, is measured in units of revolutions per minute (rpm) and may range from 10 to 2 000 rpm. Turning feed is the rate of advance of the cutting tool per revolution of the spindle. Feeds may range from 0.001 to 0.075 in (0.03 to 1.91 mm) per revolution. Considerable study has been devoted to the determination of appropriate turning speeds and feeds, with the result that there now are tables available in manufacturer's catalogs and in various machining handbooks which list recommended speeds and feeds for practically any classification of work. Such tables usually relate such variables as workpiece hardness, use of coolant, type of operation, and tool-bit material. In some plants selection of speed and feed is often heavily influenced by experience with similar parts or from previous production runs.

Effect of Speed and Feed on Cost: Speeds and feeds that are too low consume excessive time, which usually results in an increase in workpiece costs. However, optimum speeds and feeds are not necessarily the maximum that the workpiece and the machine will tolerate. Excessively high speeds and feeds result in shorter tool life and therefore in increased tool cost.

Depth of Cut

Depth of cut is a term that denotes the distance, in thousands of an inch, or in some convenient metric distance, to which the tool bit enters the work. The thickness of the metal chip thus removed is equivalent to the depth of cut. The depth of cut may vary depending upon such factors as rigidity of the particular type of lathe, the design of the tool bit, available horsepower, and the configuration and material of the workpiece. An in-feed of 0.005 in (0.13 mm), for example, results in a size reduction on a workpiece diameter of twice this amount, or 0.010 in (0.25 mm).

Tool Bits

Most metal lathe-turning operations employ single-point cutting tool bits which are ground to cut in only one direction. For proper cutting action to take place, the cutting edge must contact the workpiece before any other parts of the tool bit do. Fig. 21-6 illustrates common shapes of single-point cutting tools which are produced by grinding one end of a solid bar of tungsten or molybdenum high-speed steel.

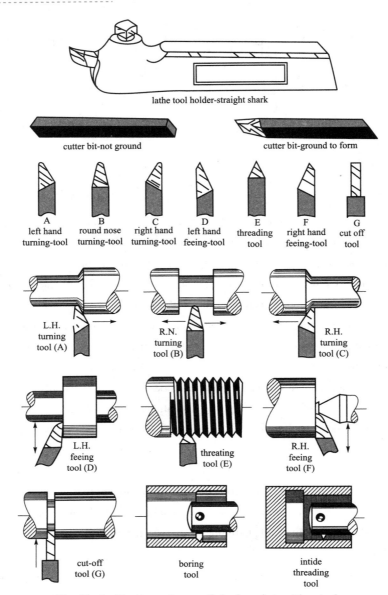

Fig. 21-6　Common shapes of single point cutting tools

Words and Expressions

axis	*n.*	轴
conical	*adj.*	圆锥的，圆锥形的
cylindrical	*adj.*	圆柱体的；圆筒形的
document	*vt.*	用文件证明
external	*adj.*	外部的；客观的

Chapter 6　Basic and Special Machining

generate	vt.	产生，发生
improvement	n.	增进，改善，改良
internal	adj.	内在的；国内的
rotation	n.	旋转
screw	n.	螺丝钉；螺旋
	vt.	调节；旋
surface	n.	表面；外表；水面
turning	n.	旋转，转向；转弯处；镟制［车削］工艺；镟坯；［复数］镟屑

1. Turning may be defined as a machining process by which cylindrical, conical, or irregularly shaped external or internal surfaces are produced on a rotating workpiece.

译文：车削可以被定义成一个机械加工过程，通过这个过程，圆柱形、圆锥形或不规则的内或外表面在一个旋转着的工件上被加工出来。

2. Speed is a term that describes the cutting speed or velocity of the rotating workpiece as it moves past the cutting edge of the tool.

译文：切削速度是用来描述当旋转的加工件通过工具切削刃口时的快慢程度的术语。

3. Effect of Speed and Feed on Cost：Speeds and feeds that are too low consume excessive time, which usually results in an increase in workpiece costs.

译文：切削速度和进给量对加工费用的影响：太低的切削速度和进给量会消耗过多的时间，通常会造成工件加工费用的增加。

4. The thickness of the metal chip thus removed is equivalent to the depth of cut.

译文：这样，被切割掉的金属屑的厚度等于切削深度。

5. Most metal lathe-turning operations employ single-point cutting tool bits which are ground to cut in only one direction.

译文：大多数的车床切削操作采用单刃工具，这一刀具进行过刃磨，只能进行单向切削。

（一）课文导读

大多数机床由两个或多个部件构成，大量生产的特种机床是由一些基本构件组成的。本章试对这些构件进行阐述。

（二）课文参考译文

机床的基本结构

大多数机床由两个或多个部件构成。虽然这些部件在加工设备如车床、铣床或钻床中有

175

着不同的功能，但是它们都有着共同的特征。

由于对去除金属的设备如车床、加工中心、铣床和磨床，以及本书中出现的其他设备的需求，现在柔性加工中心得到了持续发展。大量生产的和特殊的机床是由一些基本构件组成的。本章对这些要素进行阐述。

对机床结构的重要要求包括：刚度、形状、操作者和工件的易接近性、清除切屑简易性，以及安全性。对于机床的性能，静态和动态刚度对其精确度和精密度的影响很大。还要求机床结构的稳定性以防止加工震颤。了解基本的机床构件就能理解现代加工方法在产品制造中的宽广度。

切削机床的结构

铸件、锻件以及热成形、冷成形件通常需要切削加工。由于尺寸、形状和材料不同，因此切屑加工方式也要求多样化。

机床不仅在拥有的刃具的数量上不同，而且在工具和工件的相对运动方式上也有所不同。有一些机床（立式加工中心、钻床、钻孔机、铣床、牛头刨床和磨床）中，工件几乎保持静止而刃具运动。在其他机床（龙门刨床、车床和镗床）中，刀具几乎不动而工件运动。但应该指出的是，这些简单部件很少有不经过修改而使用的。传统的切削加工工艺见图21-1。

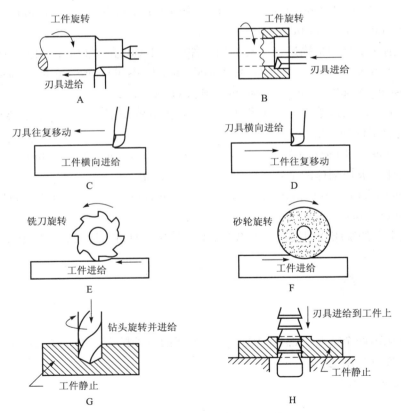

图21-1　传统的切削加工工艺
A. 车削　B. 镗削　C. 牛头刨削　D. 龙门刨削　E. 铣削　F. 磨削　G. 钻孔　H. 拉削

Chapter 6　Basic and Special Machining

单刃刀具成形的设备最直观。车床和镗床是使用单刃刀具的运动倒置设备。如图21-2A所示，在车床中工件旋转，但车刀是静止的。在镗床（图21-2B）中刀具旋转而工件静止。虽然车刀和镗床工作台并非真正静止，但暂时可以忽略。经过旋转的工件进给刀具通常比经过静止的刀架进给带有主轴箱和支座的旋转工件更为可行。牛头刨床和龙门刨床使用单刃切削刀具。图21-2C和图21-2D指出，工件尺寸是机床结构设计的一个参数。小工件在牛头刨床上比在龙门刨床上加工效率更高。转换工件和刀具的运动学关系会改变机床的外观，但是，切削动作的原理是相同的。

19世纪早期Eli Whitney引进铣的时候，这种旋转的铣刀仅仅作为镗刀使用。但Whitney开辟了铣刀新的用途。铣刀不再是仅仅用于加工圆孔，而且还用来加工键槽、开缝、锯削、平面铣和端铣、切齿以及成形不规则工件。铣床中使用了结合有工件横移的旋转刀具，如图21-2E所示。

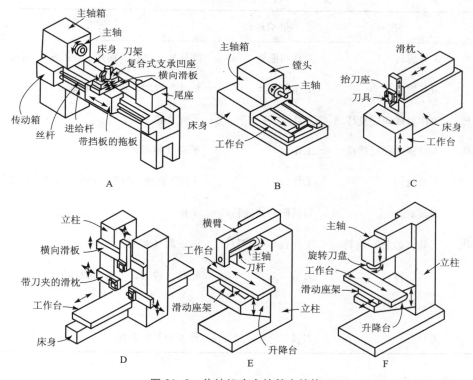

图21-2　传统机床中的基本结构
A. 车床　B. 卧式镗床　C. 牛头刨床　D. 龙门刨床　E. 卧式铣床　F. 立式铣床

标准铣床的运动倒置是落地式卧式镗床、钻床和铣床，如图21-3所示。

除了采用旋转工具（砂轮）代替单刃刀具外，外圆磨床和车床、镗床的运动方式是一样的。磨床中的工件和磨削工具按图21-4所示方式旋转。

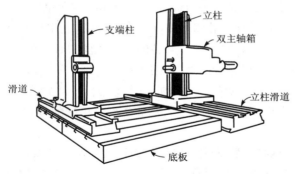 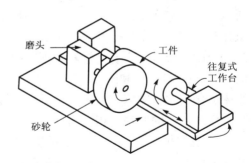

图 21-3　落地式卧式镗床、钻床和铣床的基本结构　　　　图 21-4　台式磨床的基本结构

这些切削机床的基本特征列于表 21-1 中。

表 21-1　传统机床的切削和进给运动

机床	切削运动	进给运动	操作类型
车床	工件旋转	刀具和拖板	圆柱面、钻孔、镗削、铰孔和车端面
镗床	刀具旋转	工作台	钻孔、镗削、铰孔和车端面
龙门刨床	工作台横向往复移动	刀具	平面（刨平）
牛头刨床	刀具横向往复移动	工作台	平面（整形）
卧式铣床	刀具旋转	工作台	平面、齿轮、凸轮、钻孔、镗削、铰孔和车端面
外圆磨床	刀具旋转	刀具横向往复移动	平面
卧式镗床	刀具（砂轮）旋转	工作台和/或刀具	圆柱面（磨削）
钻床	刀具旋转	刀具	钻孔、镗削、车端面和攻丝
锯床	刀具	刀具和/或工件	切断
拉床	刀具	刀具	外表面和内表面

（三）阅读材料（11）参考译文

车　　削

车削加工最常与一种被称为车床的机床联系在一起。历史上对车床的最早记载可以追溯到公元前 3 世纪的埃及。也许近年来最重要的改进是 1740 年在法国发展起来的螺纹车削原理。自那以后，这种多用途的机床不断得到稳步改进。

车削可以被定义成一个机械加工过程，通过这个过程，圆柱形、圆锥形或不规则的内或外表面在一个旋转着的工件上被加工出来。切削动作是由一个或多个与工件的旋转轴成一定角度的静止单刃车刀产生的。

工艺描述

图 21-5 说明待加工的工件表面形状取决于只有横向进给而无纵向进给的静止切削刀具的形状和大小。

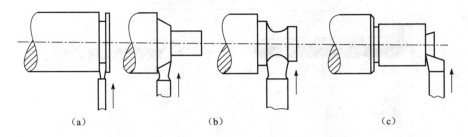

图 21-5 通过车床上的横向进给生产的典型工件表面

切削速度和进给量

切削速度和进给量是重要的切削因素。切削速度是用来描述当旋转的加工件通过刀具切削刃口时的快慢程度的术语。速度的单位是表面英尺/分钟。另一方面，转轴速度的单位是转/分钟，其范围是 10~2 000 转/分钟。车削进给量是转轴每旋转一周时车刀前进的速率。进给量的范围是每转 0.001~0.075 英寸（0.03~1.91 毫米）。已经投入了大量的研究以确定车削速度和进给量的适当值，研究在生产厂家的产品目录和各种切削加工手册中都有表可查，表中为生产实际中任何种类的工件都列出了推荐的切削速度和进给量。这些列表通常涉及一些变量，如工件硬度、冷却剂的使用、操作类型和刀具材料。在一些工厂里，切削速度和进给量的选择经常受到生产类似工件的经验或以前生产经验的重要影响。

切削速度和进给量对加工费用的影响：太低的切削速度和进给量会消耗过多的时间，这通常会造成工件加工费用的增加。然而，最佳的切削速度和进给量不必是工件和设备所能承受的最大值。过高的切削速度和进给量会缩短车刀的寿命，从而增加车刀成本。

切削深度

切削深度表示刀具进入工件的距离，以数千英寸或某种便利的米制单位来衡量。这样，被切割掉的金属屑的厚度等于切削深度。切削深度可能因一些因素如特定型号车床的刚度、刀具的设计、可利用的马力以及工件的结构和材料等而发生变化。例如，横向进给量为 0.005 英寸（0.13 毫米）时，工件直径的减小量为进给量的两倍，即 0.010 英寸（0.25 毫米）。

刀具

大多数的车床切削操作采用单刃刀具，这一刀具进行过刃磨，只能进行单向切削。为了正确切削，刀口必须在刀具的其他部件之前接触工件。图 21-6 所示为单刃刀具的常见形状，这些刀具是通过磨削钨或钼高速钢实心棒的一端而制成的。

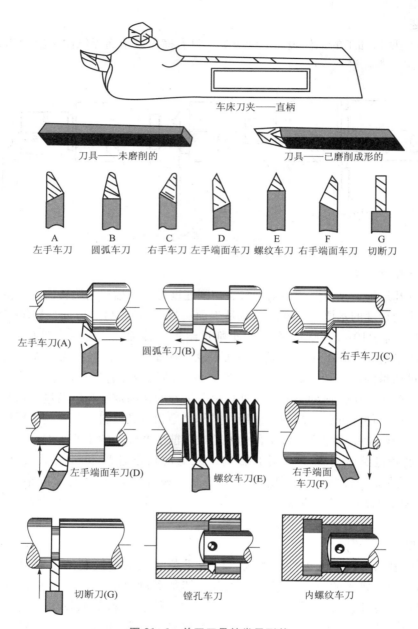

图 21-6 单刃刀具的常见形状

Lesson 22

Milling

With the possible exception of the lathe, the milling machine is regarded by most authorities as the most versatile of all machine tools. Practically all shapes and sizes of both flat and curved surfaces, on both the inside and the outside of workpieces, can usually be machined by one or another milling method. It is a uniquely adaptable and economical method of producing a wide variety of machining operations in the manufacture of only a few parts to an almost unlimited production quantity. Milling offers the additional advantage of providing methods of repetitively creating accurate parts for interchangeable manufacture.

Description of the Process

Essentially, machined surfaces are formed by the action of a rotating cutter, sometimes in a single pass of the work. The work may be held in a vise, a three-jaw chuck, an index head, a rotary table, between centers, in a special fixture, or bolted to the machine table. In most cases, the work is fed against the cutter. The speed of the cutting tool and the rate at which the workpiece travels is dependent upon the kind of material being machined. Optimum cutting efficiency is also related to machine capacity, cutter design, workpiece requirements, workholding fixture design, and other governing factors, The metal removal rate of milling machines, when compared to lathes, shapers, and planers, is usually considerably greater.

Types of Milling Operations

In most conventional milling operations, the surfaces generated by a rotating toothed cutter may be classified into two general categories, peripheral and face milling, illustrated in Fig. 22-1.

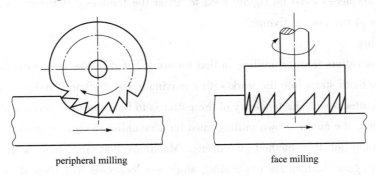

Fig. 22-1　Principle of peripheral and face milling

181

Peripheral Milling

In peripheral milling, the cutter rotates about an axis that is parallel to the surface being cut. As will be shown in subsequent illustrations, both flat and formed surfaces are produced in this way.

Face Milling

In face milling, the cutter rotates about an axis that is generally at right angles to the surface being cut. High-speed face milling with carbide-tipped cutters produce the best surface finish. Face-milling operations offer greater versatility in feeds and speeds and reduce initial cutter cost and subsequent maintenance. Face milling generally results in high production rates at the lowest cost per part.

Fig. 22-2 shows two different methods of generating surfaces by peripheral milling operations: up (or conventional) milling and down (or climb) milling.

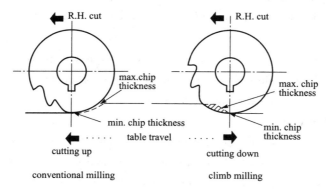

Fig. 22-2 Surfaces may be generated by up milling or down milling

Up Milling

Up milling is the condition when the work is fed against the direction of the rotating milling cutter. The chip is very thin where the tooth first contacts the work and increases in thickness to a maximum where the tooth breaks out of the work. The initial tooth contact usually occurs in clean metal and ends by lifting or peeling off the rough surface scale. Because of this effect, this is the preferred method for machining sand castings, forgings, or metals that have a rough or hard abrasive surface scale. Workpieces must be rigidly held to offset the tendency of the cutting forces to lift or pull the work out of the vise or fixture.

Down Milling

Down milling differs from up milling in that the work is fed in the same direction as the rotating cutter. The cutter tooth starts into the work with a maximum cut thickness and ends with a thin chip, resulting in less cutter wear. The tendency of the cutter is to hold the work down and, in fact, to pull the workpiece under the cutter. Down milling must be accomplished only on machines that have been specifically designed for this method of cutting. Machines that are made with backlash in the leadscrew or feeding mechanism are unsuitable, since any looseness will allow the cutter to draw the workpiece ahead and take bites that are too large. Down milling usually produces a better surface finish on harder steels than does up milling. Chips are disposed of more readily and are less likely to be carried along by the teeth. It is the method that is usually selected for milling operations on slender and intricate parts. Fixture design is simplified. Less power is required. Increased cutting speeds and feeds are practical.

Words and Expressions

authority [ɔːˈθɔriti]	n.	权威，威信；权威人士；权力，职权
backlash [ˈbæklæʃ]	n.	反斜线(\)；后座；后冲
bolt [bəult]	n.	螺钉；门闩
essentially [iˈsenʃəli]	adv.	本质上，本来
repetitive [riˈpetitiv]	adj.	重复的，反复性的
slender [ˈslendə]	adj.	苗条的；微薄的；微弱的
three-jaw chuck		三爪卡盘

Notes

1. Practically all shapes and sizes of both flat and curved surfaces, on both the inside and the outside of workpieces, can usually be machined by one or another milling method.

译文：实际上各种形状和尺寸的平面和曲面工件的内部和外部都可以由相应的铣削工艺来加工。

2. Optimum cutting efficiency is also related to machine capacity, cutter design, workpiece requirements, workholding fixture design, and other governing factors.

译文：最佳的切削效率也与机器功率、刀具设计、工件需求、夹具设计和其他控制因素有关。

3. The metal removal rate of milling machines, when compared to lathes, shapers, and planers, is usually considerably greater.

译文：与车床、牛头刨床和龙门刨床相比，铣床的金属切削率通常要大得多。

解析：compare to：与……比较。when引导过去分词短语作时间状语。

4. Face-milling operations offer greater versatility in feeds and speeds and reduce initial cutter cost and subsequent maintenance.

译文：端面铣削加工为进给量和进给速度提供了更多选择，从而减少了初始的刀具成本和以后的维护工作。

5. Down milling differs from up milling in that the work is fed in the same direction as the rotating cutter.

译文：顺铣加工与逆铣加工的不同在于，其工件的进给方向与盘型铣刀（在切削点）的旋转（的切线）方向相同。

解析：differ from：不同于，异于。

6. The cutter tooth starts into the work with a maximum cut thickness and ends with a thin chip, resulting in less cutter wear.

译文：铣刀齿一开始就切到最大深度，最后铣出的是薄切屑，这样铣刀的磨损较小。

解析：result in：引致，导致。

Exercise

Fill in the blanks with the proper words.

1. In most cases, the work _____ against the cutter. The speed of the cutting tool and the rate at which the workpiece travels _____ the kind of material being machined.
2. Optimum cutting efficiency is also related to machine capacity, cutter design, workpiece requirements, workholding fixture design, and other governing factors, The metal removal rate of milling machines, when _____ to lathes, shapers, and planers, is usually _____ greater.
3. Face-milling operations offer greater _____ in feeds and speeds and reduce initial _____ cost and _____ maintenance.

Reading Materials (12)

Section A   Grinding

Precision grinding and rough grinding each consist of forming surfaces by the use of a rotating abrasive wheel composed of many small and hard bonded abrasive grains. Each individual and irregularly shaped grain acts as a cutting tool. Rough grinding is a commonly used method for removing excess material from castings, forgings, and weldments, or as a method of removing or snagging thin fins, sharp, corners, burrs, or other unwanted projections from various shapes of workpieces. Small parts are often hand-held and moved into contact with a rotating abrasive wheel. Surfaces on larger work are ground by manually moving a portable abrasive tool over the workpiece surfaces. This chapter will deal only with precision grinding processes.

Description of the Process

Precision grinding is the principal production method of cutting materials that are too hard to cut by other conventional tools or for producing surfaces on parts to tolerance or finish requirements more exacting than can be achieved by other manufacturing methods. Surfaces on workpieces are simultaneously produced by grinding to an accurate size and with a superior surface finish. The process of precision grinding is principally associated with the removal of small amounts of material to close tolerances and should not be confused with the process of abrasive machining. Abrasive machining relates to the rapid removal of relatively larger amounts of stock, with the added capability of special workpiece shape formation. As compared with precision grinding, abrasive machining requires heavier machines with increased power capacities.

Grinding Wheel Shapes

Grinding wheel shapes have also been standardized by manufacturers. Some of the commonly used types are shown in Fig. 22-3. A variety of standard face contours for straight grinding wheels are shown in Fig. 22-4.

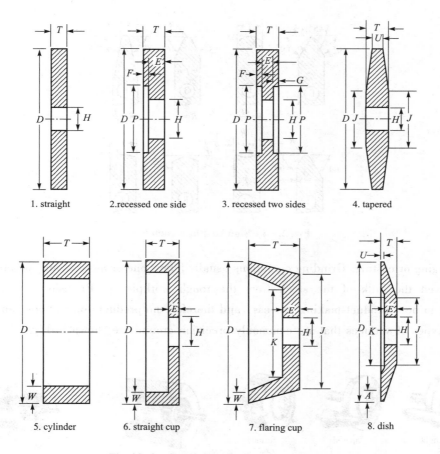

Fig. 22-3 Standard grinding wheel shapes

Typical applications of various grinding operations are shown in Fig. 22-5.

Product Applications

Other parts will explain how grinding may, in some cases, supplant planing, shaping, turning, or milling as a method for removing as much as 0.025 in (0.64 mm) of stock in one pass of the grinding wheel. Precision grinding is a term that is restricted to the method of removing small amounts of material, typically 0.001 in (0.03 mm) per pass, for improving dimensional accuracy and surface smoothness. Grinding is not generally considered to be a primary forming process in the manufacture of parts. Precision grinding operations are performed on workpieces that have been previously rough-shaped by some other primary forming process

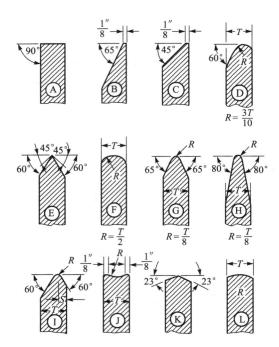

Fig. 22-4 Standard face contours

such as forging or casting. Grinding operations usually follow one or more sizing operations which have removed the bulk of the stock from the rough workpieces. Precision grinding is not considered to be a product-making process, and therefore no product applications can be listed. The main types of surfaces that are commonly precision-ground are shown in Fig. 22-5.

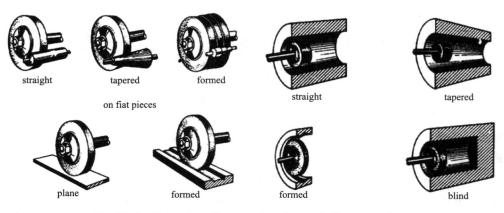

Fig. 22-5 Typical application of various grinding operations

Chapter 6　Basic and Special Machining

New Words and Expressions

abrasive	*n.*	研磨剂
	adj.	研磨的
weldment	*n.*	焊(接)件，焊成件
snag	*n.*	障碍
	v.	阻碍
contour	*n.*	轮廓；周线；等高线
dimensional	*adj.*	空间的

Notes

1. Precision grinding and rough grinding each consist of forming surfaces by the use of a rotating abrasive wheel composed of many small and hard bonded abrasive grains.

译文：精磨和粗磨都是利用由许多小而硬的磨粒黏结成的砂轮的旋转对工件表面进行加工。

2. Precision grinding is the principal production method of cutting materials that are too hard to cut by other conventional tools or for producing surfaces on parts to tolerance or finish requirements more exacting than can be achieved by other manufacturing methods.

译文：精密磨削是加工传统切削工具难以加工的硬质材料的主要方法，这种加工方式所能达到的公差和表面粗糙度要高于其他加工方法。

3. As compared with precision grinding, abrasive machining requires heavier machines with increased power capacities.

译文：与精磨相比，粗磨加工需要的是具有大功率的较大的机器。

4. Grinding wheel shapes have also been standardized by manufacturers.

译文：磨轮形状也由制造商定标准化了。

5. Precision grinding operations are performed on workpieces that have been previously rough-shaped by some other primary forming process such as forging or casting.

译文：精密磨削加工对经过诸如锻造或铸造后的一些粗加工成形的工件进行加工。

Section B　Machining Process Selection Factors

Materials：Any material that can be machined by any other method can also be ground. All hardened or soft ferrous metals, as well as all the nonferrous metals, can be ground, but some are more difficult to grind than others. Nonmetallics that are difficult or impossible to machine by conventional methods can be ground. These include ceramics, tile, marble, glass, rubber, plastics, and a wide variety of composite materials. There are practically no limitations to the kinds of materials that cad be ground.

Tolerances: The dimensional accuracy of surfaces that are sized and finished by precision grinding can vary widely depending upon the type of operation, grinding wheel used, use of a grinding fluid, and the grinding procedure (regarding the amount of stock removal). As a general rule, tolerances as small as ±0.000 1 in (0.003 mm) may be regularly obtained by production grinding methods. Special size-control devices on grinding machines are available that will allow grinding to tolerances as small as ±0.000 025 in (0.000 6 mm). As in all other machining processes, close tolerances increase the cost of grinding.

Surface Finish: Surface finishes as low as 5 to 15 μin may be regularly obtained by general production grinding. The cost of obtaining the required surface finish varies with the type of grinding wheel and equipment, the number of dressings required, workpiece hardness, and other important factors, including feed rate and in-feed per pass.

Words and Expressions

devices	n.	装置；设计；图案；策略；发明物；设备
hardened	adj.	变硬的；坚毅的；
nonmetallics	adj.	非金属的
	n.	非金属物质

1. Nonmetallics that are difficult or impossible to machine by conventional methods can be ground.

译文：传统方式难以加工的非金属物质也能被磨削。

2. Tolerances: The dimensional accuracy of surfaces that are sized and finished by precision grinding can vary widely depending upon the type of operation, grinding wheel used, use of a grinding fluid, and the grinding procedure (regarding the amount of stock removal).

译文：公差：由精密研磨而成的表平面的尺寸精度会随着加工的类型、使用的砂轮、研磨液的使用与否以及研磨的程序而不断变化（与材料的去除量有关）。

3. Surface finishes as low as 5 to 15 μin may be regularly obtained by general production grinding.

译文：普通的磨削加工可以得到5~15微英寸的表面光洁度。

（一）课文导读

铣床是适用性最广泛的机械工具，本文论述了铣削的过程和铣削加工的类型。

（二）课文参考译文

铣 削

除车床外，大部分权威人士认为在所有机床种类中铣床的适用性最广。实际上各种形状和尺寸的平面和曲面工件的内部和外部都可以由相应的铣削工艺来加工。这是一种少有的适应性强且经济实惠的工艺方法，因为它在制造几个到几乎无限个工件的过程中能进行许多不同的加工操作。铣削的另一个优势在于它也是一种为互换性生产重复制造精密工件的方法。

工艺描述

从本质上来说，加工面是通过旋转铣刀加工而成的，有时是在工件的单行程中。工件被夹持在台虎钳、三爪卡盘、分度头、分度台、轴间、专用夹具中，或者固定在工作台上。大多数情况下，工件的进给方向与铣刀的相反。铣刀的速度和工件移动的速率取决于加工的材料种类。最佳的切削效率也与机器功率、刀具设计、工件需求、夹具设计和其他控制因素有关。与车床、牛头刨床和龙门刨床相比，铣床的金属切削率通常要大得多。

铣削加工的类型

在最传统的铣削加工中，旋转的齿形铣刀加工出的平面通常分为两大类：圆周铣削和端面铣削，如图22-1所示。

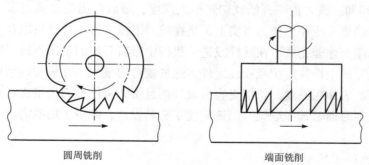

圆周铣削　　　　　　　　端面铣削

图22-1　圆周铣削和端面铣削的原理

圆周铣削

在圆周铣削中，铣刀围绕一根与加工面平行的轴旋转。如图22-2所示，平面或成型面都是以这种方式加工的。

端面铣削

在端面铣削中，铣刀围绕一根与加工面垂直的轴旋转。用硬质合金铣刀进行高速端面铣削可以获得最好的表面光洁度。端面铣削加工为进给量和进给速度提供了更多选择，从而减少了初始的刀具成本和以后的维护工作。一般说来，端面铣削能以最低的单个工件成本获得高生产率。

图22-2所示为通过圆周铣削加工表面的两种不同方法：逆铣（或惯用铣削）和顺铣（或同向铣削）。

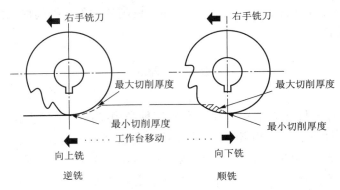

图22-2 逆铣或顺铣加工的表面

逆铣

当铣刀的进给方向与旋转铣刀的进给方向相反时，称为逆铣。在铣刀齿与工件初始接触的地方切屑非常薄，然后切屑厚度逐渐增加，在铣刀齿切断工件的地方切屑厚度最大。刀齿通常最先接触洁净的金属，最后与粗糙的表层脱离接触。因为逆铣具有这种效果，所以对于加工砂型铸件、锻件，或者具有粗糙或坚硬的耐磨表面而言，逆铣是一种较好的方法。工件必须被紧紧地夹住，以防止在切削刀的作用下被从虎钳或夹具上顶起或拉出。

顺铣

顺铣加工与逆铣加工的不同在于，其工件的进给方向与盘型铣刀（在切削点）的旋转（的切线）方向相同。铣刀齿一开始就切到最大深度，最后铣出的是薄切屑，这样铣刀的磨损较小。铣刀的趋势是压住工件，事实上，是在铣刀下拉工件。顺铣只能在为这种切削方法而专门设计的机床上才能完成。在丝杆或进给机构中带有后座的机床不适于顺铣，因为任何松动都将会使铣刀把工件往前拉或咬住工件，这种咬合力很大。在较硬的钢件上顺铣通常比逆铣能获得更好的表面粗糙度。切屑更容易处理而且不太可能被刀齿带走。在细长的和复杂的工件上进行铣削时通常选择顺铣。顺铣简化了夹具设计，减少了能源需求，并且增大了切削速度和进给量。

（三）阅读材料（12）参考译文

A 磨 削

精磨和粗磨都是利用许多小而硬的磨粒黏结成的砂轮的旋转对工件表面进行加工。每一个单独的、形状不规则的磨粒相当于一个切削刀具。粗磨是一种去除铸件、锻件和焊接件上多余材料的常用方法，也是一种去除或打磨各种形状的工件上的薄飞翅、锐边、棱角、毛边或其他不想要的突起部分的方法。小工件通常是用手拿着并移送去与旋转的砂轮接触。较大工件的表面是通过在工件表面上手工移动手提式砂轮来进行磨削的。本部分只就精磨工艺进行阐述。

工艺描述

精密磨削是加工传统切削工具难以加工的硬质材料的主要方法，这种加工方式所能达到的公差或表面粗糙度要高于其他加工方法。工件表面通过精磨可以同时达到精确的尺寸和优异的表面粗糙度。精磨主要去除少量材料以达到精确公差，不应该将其与粗磨混淆。粗磨是指快速去除相对大量的坯料，还有使特殊工件成形的附加功能。与精磨相比，粗磨加工需要的是具有大功率的较大的机器。

砂轮形状

砂轮形状也已经由制造商标准化了。一些常用的砂轮类型如图 22-3 所示。各种砂轮的标准端面轮廓如图 22-4 所示。不同磨削操作的典型应用示于图 22-5 中。

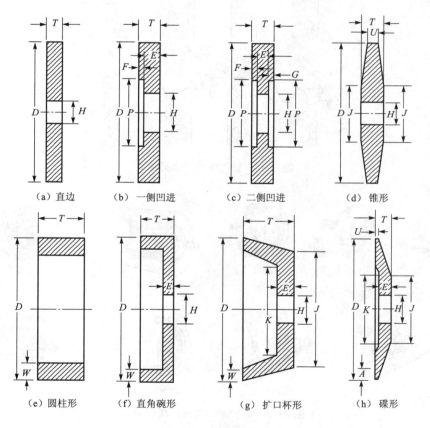

图 22-3　标准的砂轮形状

产品应用

其他章节会阐述在某些情况下磨削作为一种一次磨削量为 0.025 英寸（0.64 毫米）的方法是如何来取代龙门刨削、牛头刨削、车削或铣削的。精磨是指为了提高尺寸精度和表面光洁度而去除少量的材料的方法，尤指每次磨削量为 0.001 英寸（0.03 毫米）。磨削通常不是工件制造过程中的初始工艺。精密磨削加工对经过诸如锻造或铸造后的一些粗加工成形的工件进行加工。磨削加工通常是在一次或多次定尺寸操作之后，定尺寸操作已经把粗糙工件上的大部分坯料去除掉了。精磨不是一种制造产品的工艺，因此不能列举出具体的产品应用。通常由精磨加工而成的主要的表面类型如图 22-5 所示。

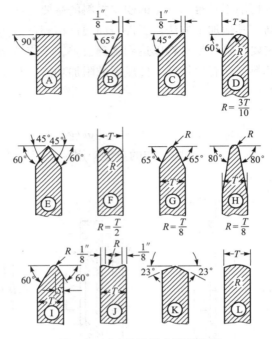

图 22-4　砂轮的标准端面轮廓

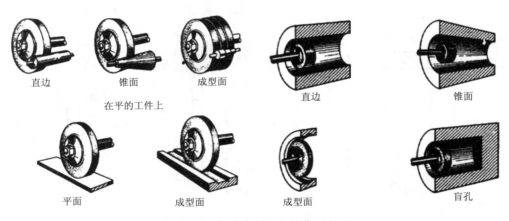

图 22-5　各种磨削工艺的典型应用

B　磨削加工的选择因素

材料：能被其他方法切削加工的任何材料都能被磨削。所有硬的或软的黑色金属，以及所有的有色金属都能被研磨，但是它们的磨削难易程度不一样。传统方式难以加工的非金属物质也能被磨削。这些物质包括陶瓷、瓷瓦、大理石、玻璃、橡胶、塑料以及各种各样的复合材料。几乎所有的物质都能进行磨削。

公差：由精密研磨而成的表平面的尺寸精度会随着加工的类型、使用的砂轮、研磨液的使用与否以及研磨的程序而不断变化（与材料的去除量有关）。作为一般规律，磨削可以达到的尺寸公差是±0.0001英寸（0.003毫米）。磨床上的特殊尺寸控制装置可以使磨削达到±0.000 025英寸（0.0006毫米）的公差。正如其他所有的切削工艺一样，精确的公差会增加磨削成本。

表面光洁度：普通的磨削加工可以得到5~15微英寸的表面光洁度。达到要求的表面光洁度所需要的成本根据砂轮和设备的类型、要求的磨削件数量、工件的硬度以及进给速率和每次进给量等其他一些重要因素而定。

Lesson 23

Electrical Discharge Machining

Product Applications

Electrical Discharge Machining is the process normally referred to as EDM. With the increasing use of hard, difficult-to-machine space-age metals, EDM's ability to machine burr-free, intricate configurations, narrow slots, and blind cavities or holes into these materials, at close tolerance, becomes more important. Since EDM does not set up the high cutting forces and mechanical strains often associated with conventional machining, the process is well suited for cutting tubing, honeycomb, and other thin-wall, fragile structures. Small hole drilling, 0.005 in (0.13 mm) in diameter and as deep as 20 diameters, with virtually no bending or drifting, for example, is possible. For comparison, in conventional hole drilling, when the length of a twist drill is 10 times the hole diameter or greater, it is difficult to hold straightness. Slots as narrow as 0.001 in (0.03 mm), recesses, reentrant features, holes in awkward locations, and even curved holes may be produced by this remarkable process.

One well-established application of EDM is in machining die cavities and molds used for die casting, plastic molding, wiredrawing, extrusion, compacting, cold-heading, and forging. In this widespread area of application, the EDM process is known as "die sinking."

Another important application of EDM is in the metal-forming field to produce punch, trim, or stamping dies. This category of work requires that a precisely sized through hole be made, often irregular in shape, in a solid carbide die. Electric discharge machining is ideally suited to the manufacture of dies of this type, as well as for progressive or compound dies, which are a combination of through hole and cavity dies.

Another variation of EDM uses a wire traveling from reel to reel instead of using a shaped tool for the electrode. The schematic diagram shown in Fig. 23-1 illustrates the principle of a moving-wire electrode attachment. The wire is gradually advanced between the reels to compensate for the wear that occurs at the point of cutting. This innovative variation in the EDM process amounts to what is essentially an "electronic bandsaw," which cuts precision slots by moving at a constant speed across the workpiece. Initially, this type of equipment was used as little more than a slicing maching for tubing, thin-walled structures, and honeycombing. Now, some companies have added numerical control-positioning features so that complex two-dimensional shapes can be cut. EDM can thus be put to work on simple profiling jobs without the benefit of special electrodes. It is often possible to stack the workpieces in cases where the relatively slow cutting speed of 1 in²/h (6.3 cm²/h) is objectionable on an otherwise one-piece basis.

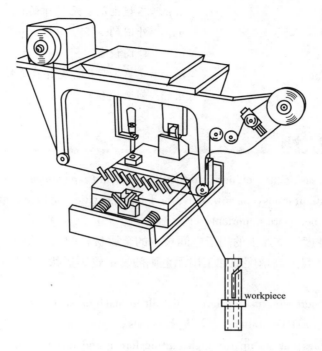

Fig. 23-1 Schematic diagram showing the principle of a moving wire electrode attachment used to simultaneously cut a slot on ten identical workpieces shown in the enlarged inset (Courtesy Easco-Sparcatron, Inc)

Another ingenious design feature consists of a rotating electrode in the form of a tube, disk, or other special shape. Somewhat analogous to grinding, the rotating electrode has the advantage of a large effective surface area. As a consequence, the wear on the electrode is considerably less than that for conventional electrodes. Electrodes such as these can be mounted on the machine in the peripheral or planer mode.

Words and Expressions

analogous [əˈnæləgs]	adj.	类似的，相似的，可比拟的
attachment [əˈtætʃmənt]	n.	附件，附加装置，配属
carbide [ˈkɑːbaid]	n.	[化]碳化物
cavity dies	n.	型腔模
configuration [kənˌfigjuˈreiʃən]	n.	构造，结构；配置；外形
electrode [iˈlektrəud]	n.	电极

honeycomb [ˈhʌnikəum]	n.	蜂房；蜂巢；蜂脾
intricate [ˈintrikit]	adj.	错综复杂的，难以理解的
irregular [iˈregjulə]	adj.	不规则的，无规律的
peripheral [pəˈrifərəl]	adj.	外围的
	n.	外围设备
variation [ˌvɛəriˈeiʃən]	n.	变更，变化，变异；变种；[音]变奏，变调
wiredrawing	n.	制造铁丝；拖长

Notes

1. With the increasing use of hard, difficult-to-machine space-age metals, EDM's ability to machine intricate configurations, narrow slots, and blind cavities or holes into these materials, at close tolerance, becomes more important.

译文：随着高硬度、难加工的用于空间技术的金属材料的大量使用，电火花加工具有的这种易于在难加工材料上高精度地加工出无毛刺的复杂轮廓形状、窄槽、腔体以及盲孔的能力变得更加重要。

解析：with the increasing use of hard, difficult-to-machine space-age metals 在句中作方式状语。此句中的主语是 EDM's ability，谓语是 becomes。

2. Since EDM does not set up the high cutting forces and mechanical strains often associated with conventional machining, the process is well suited for cutting tubing, honeycomb, and other thin-walled, fragile structures.

译文：由于电火花加工不像传统的机械加工那样产生明显的切削力和机械应力，所以这种工艺适用于加工带有管状、蜂窝结构以及其他脆性较大的薄壁件。

解析：associate with：联系。 be suited for：适用于。

3. One well-established application of EDM is in machining die cavities and molds used for die casting, plastic molding, wiredrawing, extrusion, compacting, cold-heading, and forging.

译文：EDM（电火花加工）广泛地应用于加工模腔和压铸模、塑性成形模、拉丝模、挤压模、压缩模、冷镦模和锻模。

4. Another important application of EDM is in the metal-forming field to produce punch, trim, or stamping dies.

译文：EDM（电火花加工）的另一个重要应用是在金属成形领域中制造冲裁模、切边模和冲压模。

5. Another variation of EDM uses a wire traveling from reel to reel instead of using a shaped tool for the electrode.

译文：EDM 的另一种变化是用从一个卷轴到另一卷轴运转的一根切割丝代替成形工具作为电极。

Chapter 6 Basic and Special Machining

Exercise

Fill in the blanks with the proper words.

1. Another important application of EDM is in the metal-forming field _____ punch, trim, or stamping dies. This category of work _____ that a precisely sized through hole be made, often irregular in shape, in a solid carbide die.
2. Electric discharge machining is ideally _____ to the manufacture of dies of this type, as well as for progressive or compound dies, which are a _____ of through hole and cavity dies.
3. Another _____ of EDM uses a wire traveling from reel to reel instead of using a _____ tool for the electrode.
4. Another ingenious design feature _____ of a _____ electrode in the form of a tube, disk, or other _____ shape.

课文导读及参考译文

（一）课文导读

电火花加工就是指常被称为 EDM 的加工过程。本章介绍了电火花加工在模具制造中的应用。

（二）课文参考译文

电火花加工

产品应用

电火花加工就是通常被称为 EDM 的加工工艺。随着高硬度、难加工的用于空间技术的金属材料的大量使用，电火花加工具有的这种易于在难加工材料上高精度地加工出无毛刺的复杂轮廓形状、窄槽、腔体以及盲孔的能力变得更加重要。由于电火花加工不像传统的机械加工那样产生明显的切削力和机械应力，所以这种工艺适用于加工带有管状、蜂窝结构以及其他脆性较大的薄壁件。例如，钻一个直径为 0.005 英寸（0.13 毫米）、深度为 20 倍直径的小孔，没有弯曲或偏移，是完全可能的。相比而言，当麻花钻的长度是小孔直径的 10 倍或更多倍时，传统的钻孔操作难以保证钻得笔直。窄到 0.001 英寸（0.03 毫米）的缝、凹槽、凹腔、不便操作处的小孔甚至曲折的小孔都能通过此工艺加工。

EDM（电火花加工）广泛地应用于加工模腔和压铸模、塑性成形模、拉丝模、挤压模、压缩模、冷镦模和锻模。在如此广泛的应用中，EDM（电火花加工）被称为"刻模"。

EDM（电火花加工）的另一个重要应用是在金属成形领域中制造冲裁模、切边模或冲压模。这种类型的制造要求在坚固的硬质合金模中加工出形状不规则的、尺寸精确的通孔。电火花加工非常适于制造这种类型的模具，以及连续模或通孔与型腔模组合而成的复合模。

EDM 的另一种变化是用从一个卷轴到另一个卷轴运转的一根切割丝代替成形工具作为电极。图 23-1 显示了移动切割丝电极附属装置的原理。切割丝在两卷轴之间逐渐前移来补偿切割点处的磨损。EDM 工艺中的这种创新本质上相当于"电动带锯",通过恒速穿过工件来切割出精确的槽。这种设备最初用来对管状、薄壁结构和蜂窝结构进行切片加工。现在,有些公司增加了数字控制的定位装置,因此能加工出复杂的二维形状。所以 EDM 不需要特殊的电极就能进行简单的仿形加工。在切削速度为单件生产时不允许的、相对较慢的 1 平方英寸/小时(6.3 平方厘米/小时)的情况下,通常可以叠放工件。

另一个巧妙的设计是管状、盘状或其他特殊形状的旋转电极。对形状有些类似的工件进行磨削时,旋转电极具有较大的有效表面积。因此,旋转电极的磨损比传统电极的磨损要小得多。这种电极能以外围设备或龙门刨床的形式安装在机床上。

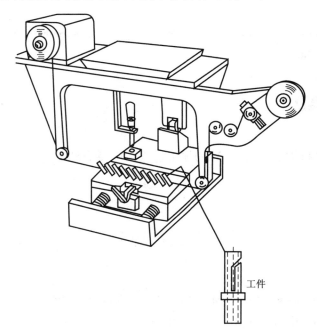

图 23-1 移动切割丝电极附属装置的原理示意图,该装置用来在 10 个相同工件
(如放大的插图中所示)上同时切割缝隙
(经 Easco-Sparcatron 公司许可)

Chapter 7

The Application of Computer in Design and Manufacture of Mould and Die
（计算机在模具设计与制造中的应用）

Lesson 24

Computers and CAD/CAM

Computers first appeared in the 1940's. The early inventions were bulky, cumbersome to use and performed computations rather slowly in comparison to modern digital computers. They advanced from mechanical relays to vacuum tubes to transistors to silicon chips. Computers today are more compact, faster and less expensive than their predecessors. Today some microcomputers can handle the computations of early day main-frames. Technological advances in this field have been truly dramatic.

The application areas of computers have also grown rapidly. Computers are widely used in the fields of engineering, business, education and medicine. However, the most spectacular growth has been in the area of CAD/CAM — Computer Aided Design and Computer Aided Manufacture. This new technology, which emerged within the last decade or so, has helped to increase engineering productivity tremendously. Higher productivity is probably the primary consideration that influences most potential users to acquire a CAD/CAM system. CAD/CAM provides the integration of design, analysis and manufacturing functions into a system which is available to the user at his fingertips. In addition, other routine and monotonous (but important) tasks such as the preparation of bills of materials, costing, production scheduling, etc. may be performed automatically using the same computer network. Another major benefit in the use of CAD/CAM is reduced lead time from concept to design to manufacture. Product development cost can also be reduced dramatically because analysis such as the finite element method can be interfaced with design to arrive at the optimum design within a very short time.

CAD/CAM and Analysis are best used in an interactive, computer graphics environment to provide solutions to engineering problems as illustrated in Fig. 24-1. The geometry and topology of the part and other parameters such as material properties are stored in a common data base which can be accessed by any one of the functions. Analysis such as finite element methods, upper bound methods and slab analysis is carried out on the initial product geometry.

Results which are often displayed in colour will either automatically or with user interventions alter the design parameters so as to satisfy the various design criteria. This process is repeated until an optimum solution is obtained. Thereafter manu-facturing and other related functions can be performed either automatically or with user intervention.

Any CAD/CAM system consists of hardware, such as the central processing unit (CPU), disk

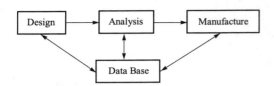

Fig. 24-1 Interactive computer graphics environment

storage facilities, display units, tablets, etc. and software, which is the brain behind the system. Sometimes it is harder to make a clear distinction between the two, when hardware is built-in with appropriate software. This is often referred to as "firmware."

A typical CAD/CAM system configuration is illustrated in Fig. 24-2. The heart of the system is the central processing unit (or the computer), which coordinates all the functions within the CAD/CAM system. The user interacts with the system through the work station. A typical work station is shown in Fig. 24-3. It consists of a graphic display (or two, as in this case) which provides visual output of the system to the user. The user can communicate with the system through the keyboard or a tablet (with menu). The dual-screen workstation has numerous advantages. It can be used to view the product being designed at two or more angles. The zoom command may be used to zoom up a certain feature of the product on one screen for detail view while the other screen can be used to view the whole part. Text may be entered on any one of the screens. Other vendors have systems with graphic display and an alpha-numeric display for text.

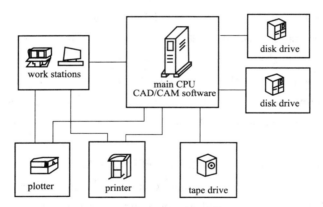

Fig. 24-2 Typical CAD/CAM system configuration

The work stations are intelligent, in that they do useful local operations such as zoom, rotate, etc, without tying up the CPU for these operations. The current trend is to increase the power of the work station, thus freeing the CPU for other more demanding tasks. Many CAD/CAM systems have a "menu" of commands to choose from a tablet or digitizer board, The menus can be changed depending upon the applications. Some CAD/CAM systems use a push-button menu to select certain commands. A typical menu used in a CAD/CAM system is shown in Fig. 24-3.

Chapter 7 The Application of Computer in Design and Manufacture of Mould and Die

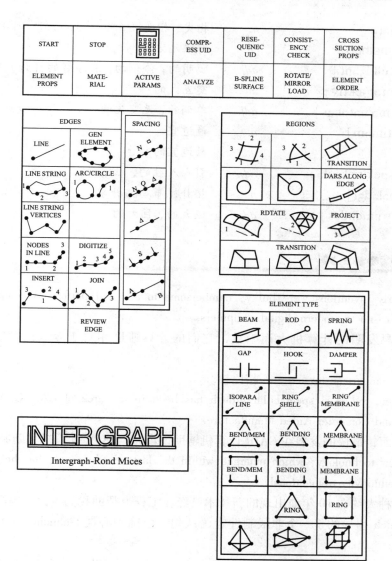

Fig. 24-3 Typical CAD/CAM menu (Courtesy of Intergraph Corp.)

 Words and Expressions

bulky [ˈbʌlki]	adj.	大的，容量大的，体积大的
coordinate [kəuˈɔːdinit]	n.	同等者，同等物；坐标（用复数）
	adj.	同等的，并列的
	vt.	调整，整理
cumbersome [ˈkʌmbəsəm]	adj.	讨厌的，麻烦的，笨重的
dramatic [drəˈmætik]	adj.	戏剧性的，生动的

fingertip ['fiŋgətip]	n.	指尖；指套
finite element method		有限元方法
intelligent [in'telidʒnt]	adj.	聪明的，伶俐的，有才智的 [计] 智能的
interactive [ˌintər'æktiv]	adj.	交互式的
monotonous [mə'nɔtənəs]	adj.	单调的，无变化的
optimum ['ɔptiməm]	n.	最适宜
	adj.	最适宜的
spectacular [spek'tækjulə]	adj.	引人入胜的，壮观的
topology [tə'pɔlədʒi]	n.	拓扑；布局拓扑学
tremendous [tri'mendəs]	adj.	极大的，巨大的

Notes

1. The early inventions were bulky, cumbersome to use and performed computations rather slowly in comparison to modern digital computers.

译文：最早发明的计算机大而笨重，它们的计算速度相对于今天的数字计算机要慢得多。

解析：in comparison to：与……相比较。

2. However, the most spectacular growth has been in the area of CAD/CAM — Computer Aided Design and Computer Aided Manufacture.

译文：然而，最快的发展在CAD(计算机辅助设计)和CAM(计算机辅助制造)领域。

3. This new technology, which emerged within the last decade or so, has helped to increase engineering productivity tremendously.

译文：这种大约在10年前出现的新技术已经在工程上帮助我们极大地提高了生产力。

解析：which 引导的是一个非限制性定语从句，修饰中心词 technology；句中的 or so 意为"大约"。

4. CAD/CAM provides the integration of design, analysis and manufacturing functions into a system which is available to the user at his fingertips.

译文：CAD/CAM 为用户提供了一个集设计、分析、制造功能于一体的系统，用户只需动动手指就能使用。

解析：provide into：提供。be available to：可用的，可得到的。

5. In addition, other routine and monotonous (but important) tasks such as the preparation of bills of materials, costing, production scheduling, etc. may be performed automatically using the same computer network.

译文：此外，其他一些日常的单调的（但是重要的）任务，比如说材料清单的准备、成本计算、生产调度等也许可以通过使用同一电脑网络自动执行。

6. Analysis such as finite element methods, upper bound methods and slab analysis is carried out on the initial product geometry.

译文：像有限元法、上限元法和板坯分析法等分析都在初始的产品几何形状上进行。

Chapter 7　The Application of Computer in Design and Manufacture of Mould and Die

解析：carry out：执行。

7. Results which are often displayed in colour will either automatically or with user interventions alter the design parameters so as to satisfy the various design criteria, This process is repeated until an optimum solution is obtained.

译文：通常用彩色显示出来的结果既可以自动也可以通过用户的介入来改变设计参数，以满足各种设计标准，这个过程可以不断地重复下去，直到获得最优的解决方案为止。

8. Any CAD/CAM system consists of hardware, such as the central processing unit (CPU), disk storage facilities, display units, tablets, etc. and software, which is the brain behind the system.

译文：任何 CAD/CAM 系统都包括硬件——如中央处理器（CPU）、磁盘存储器、显示器、写字板等等，以及软件——它是整个系统的大脑。

9. The heart of the system is the central processing unit (or the computer), which coordinates all the functions within the CAD/CAM system.

译文：系统的核心是中央处理器，它可以协调 CAD/CAM 系统中的各项功能。

10. The user interacts with the system through the work station that consists of a graphic display (or two, as in this case) which provides visual output of the system to the user.

译文：用户通过工作站与系统交互作用，工作站由一个给用户提供可视化输出的图像显示器（或两个，如图）所构成。

解析：interact with =（of two things）to have an effect on each other：两个事物间相互作用。

11. The user can communicate with the system through the keyboard or a tablet (with menu).

译文：用户可以通过键盘或写字板（带有菜单）与系统交流。

解析：communicate with：交流，沟通。

12. The work stations are intelligent, in that they do useful local operations such as zoom, rotate, etc, without tying up the CPU for these operations.

译文：工作站是智能的，因为它们可以在没有与 CPU 联系的情况下进行一些有用的局部操作，如放大、旋转等。

解析：tie up：缠住某人，使某人脱不了身。

13. Many CAD/CAM systems have a "menu" of commands to choose from a tablet or digitizer board. The menus can be changed depending upon the applications.

译文：许多 CAD/CAM 系统有一个来自写字板或数字转换板的命令"菜单"可选择。菜单可以根据应用而变化。

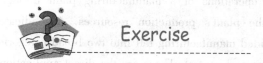

Fill in the blanks with the proper words.

1. A typical CAD/CAM system configuration _____ in Fig. 24-2. The heart of the system is the central processing unit (or the computer), which _____ all the functions within the CAD/CAM system.

2. The user interacts with the system through the work station that _____ a graphic display (or two, as in this case) which provides visual output of the system to the user.
3. The zoom command may be used to zoom up _____ feature of the product on one screen for detail view while the other screen can be used to view the whole part. Text may be entered on any one of the screens. Other vendors have systems with graphic display and an alpha-numeric display for text.

 (13)

CAD/CAM Defined

CAD/CAM is a term which means computer-aided design and computer-aided manufacturing. It is the technology concerned with the use of digital computers to perform certain functions in design and production. This technology is moving in the direction of greater integration of design and manufacturing, two activities which have traditionally been treated as distinct and separate functions in a production firm. Ultimately, CAD/CAM will provide the technology base for the computer-integrated factory of the future.

Computer-aided design (CAD) can be defined as the use of computer systems to assist in the creation, modification, analysis, or optimization of a design. The computer systems consist of the hardware and software to perform the specialized design functions required by the particular user firm. The CAD hardware typically includes the computer, one or more graphics display terminals, keyboards, and other peripheral equipment. The CAD software consists of the computer programs to implement computer graphics on the system plus application programs to facilitate the engineering functions of the user company. Examples of these application, programs include stress strain analysis of components, dynamic response of mechanisms, heat-transfer calculations, and numerical control part programming. The collection of application programs will vary from one user firm to the next because their product lines, manufacturing processes, and customer markets are different. These factors give rise to differences in CAD system requirements.

Computer-aided manufacturing (CAM) can be defined as the use of computer systems to plan, manage, and control the operations of a manufacturing plant through either direct or indirect computer interface with the plant's production resources. As indicated by the definition, the applications of computer-aided manufacturing fall into two broad categories:

1. Computer monitoring and control. These are the direct applications in which the computer is connected directly to the manufacturing process for the purpose of monitoring or controlling the process.

2. Manufacturing support applications. These are the indirect applications in which the computer is used in support of the production operations in the plant, but there is no direct interface between

Chapter 7 The Application of Computer in Design and Manufacture of Mould and Die

the computer and the manufacturing process.

The distinction between the two categories is fundamental to an understanding of computer-aided manufacturing. It seems appropriate to elaborate on our brief definitions of the two types.

Computer monitoring and control can be separated into monitoring applications and control applications. Computer process monitoring involves a direct computer interface with the manufacturing process for the purpose of observing the process and associated equipment and collecting data from the process. The computer is not used to control the operation directly. The control of the process remains in the hands of human operators, who may be guided by the information compiled by the computer.

Computer process control goes one step further than monitoring by not only observing the process but also controlling it based on the observations. The distinction between monitoring and control is displayed in Fig. 24-4. With computer monitoring the flow of data between the process and the computer is in one direction only, from the process to the computer. In control, the computer interface allows for a two-way flow of data. Signals are transmitted from the process to the computer, just as in the case of computer monitoring. In addition, the computer issues command signals directly to the manufacturing process based on control algorithms contained in its software.

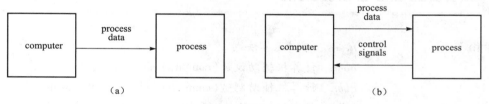

Fig. 24-4 Computer monitoring versus computer control
(a) computer monitoring (b) computer control

In addition to the applications involving a direct computer-process interface for the purpose of process monitoring and control, computer-aided manufacturing also includes indirect applications in which the computer serves a support role in the manufacturing operations of the plant. In these applications, the computer is not linked directly to the manufacturing process. Instead, the computer is used "off-line" to provide plans, schedules, forecasts, instructions, and information by which the firm's production resources can be managed more effectively. The form of the relationship between the computer and the process is represented symbolically in Fig. 24-5. Dashed lines are used to indicate that the communication and control link is an off-line connection, with human beings often required to consummate the interface. Some examples of CAM for manufacturing support that are discussed in subsequent chapters of this book include:

Numerical control part programming by computers. Control programs are prepared for automated machine tools.

Computer-automated process planning. The computer prepares a listing of the operation sequence required to process a particular product or component.

Computer-generated work standards. The computer determines the time standard for a particular production operation.

Production scheduling. The computer determines an appropriate schedule for meeting production requirements.

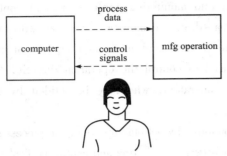

Material requirements planning. The computer is used to determine when to order raw materials and purchased components and how many should be ordered to achieve the production schedule.

Shop floor control. In this CAM application, data are collected from the factory to determine progress of the various production shop orders.

In all of these examples, human beings are presently required in the application either to provide input to the computer programs or to interpret the computer output and implement the required action.

Fig. 24-5 CAM for manufacturing support

Words and Expressions

algorithm	n.	[数]运算法则
CAD	abbr.	计算机辅助设计(computer aided design)
CAM	abbr.	计算机辅助制造(computer-aided manufacturing)
digital	adj.	数字的;数位的;手指的
	n.	数字;数字式
elaborate	adj.	精心制作的;详细阐述的;精细的
	vt.	精心制作;详细阐述
facilitate	vt.	(不以人作主语的)使容易,使便利;推动,帮助,促进
graphics	n.	(作单数用)制图法;制图学;图表算法;图形
implement	n.	工具,器具
	vt.	贯彻,实现
integrated	adj.	综合的;完整的;集成的
interface	n.	[地质]分界面,接触面;[物、化]界面
interpret	v.	解释,说明;口译,通译;认为是……的意思
terminal	n.	终点站,终端;接线端
	adj.	末期的,晚期的,每学期的

Notes

1. CAD/CAM is a term which means computer-aided design and computer-aided manufacturing.
译文:CAD/CAM 是指计算机辅助设计和计算机辅助制造。

Chapter 7 The Application of Computer in Design and Manufacture of Mould and Die

2. This technology is moving in the direction of greater integration of design and manufacturing, two activities which have traditionally been treated as distinct and separate functions in a production firm.

译文：这种技术正朝着更为综合的设计和制造的方向发展，而设计和制造在过去一直被看做是生产中两个有区别的、独立的过程。

3. Computer-aided design (CAD) can be defined as the use of computer systems to assist in the creation, modification, analysis, or optimization of a design.

译文：CAD 可以被定义为运用计算机系统对设计进行创造、修改、分析或优化。

4. Computer-aided manufacturing (CAM) can be defined as the use of computer systems to plan, manage, and control the operations of a manufacturing plant through either direct or indirect computer interface with the plant's production resources.

译文：CAM 可以被定义为运用计算机系统并结合工厂里的生产资源直接或间接地对工厂的生产过程进行计划、管理和控制。

5. Computer process monitoring involves a direct computer interface with the manufacturing process for the purpose of observing the process and associated equipment and collecting data from the process.

译文：计算机工艺监测是计算机直接地介入生产过程，以监测加工过程、相关设备和收集加工数据为目的。

6. In addition to the applications involving a direct computer-process interface for the purpose of process monitoring and control, computer-aided manufacturing also includes indirect applications in which the computer serves a support role in the manufacturing operations of the plant.

译文：除了运用计算机对生产过程直接进行监控外，计算机辅助制造还间接地对工厂的生产过程起支撑作用。

7. In all of these examples, human beings are presently required in the application either to provide input to the computer programs or to interpret the computer output and implement the required action.

译文：在所有这些例子中，人的作用是输入计算机程序或者对计算机的输出结果进行解译，以完成所需的步骤。

Exercise

Fill in the blank with appropriate forms.

In addition to the applications _____ (involve) a direct computer-process interface for the purpose of process _____ (monitor) and control, computer-aided manufacturing also includes indirect applications in which the computer serves a support role in the manufacturing operations of the plant. In these applications, the computer _____ (link) directly to the manufacturing process. Instead, the computer is used "off-line" to provide plans, schedules, forecasts, instructions, and information by which the firm's production resources can be managed more effectively. The form of the relationship between the computer and the process is represented symbolically in Fig. 24-5. Dashed lines are used to indicate that the communication and control link is an off-line connection, with human beings often required to consummate the interface.

（一）课文导读

计算机的发展日新月异。本章介绍了计算机辅助设计与计算机辅助制造在模具生产制造中的应用。

（二）课文参考译文

计算机与 CAD/CAM

第一台计算机诞生于 20 世纪 40 年代。最早发明的计算机大而笨重，它们的计算速度相对于今天的数字计算机要慢得多。它们由机械式继电器发展到真空管、晶体管再到硅芯片。与以前的计算机相比，今天的计算机设计更加紧凑，计算速度更快，价格更便宜。如今，有些微型计算机能够处理早期大型计算机所进行的计算操作。这一领域的技术进步确实是巨大的。

计算机的应用领域也快速增加。计算机被广泛应用于工程、商业、教育以及医药领域。然而，最快的发展在 CAD（计算机辅助设计）和 CAM（计算机辅助制造）领域。这种大约在 10 年前出现的新技术已经在工程上帮助我们极大地提高了生产力。影响大部分潜在用户采用 CAD/CAM 系统的主要因素可能是为了获得更高的生产力。CAD/CAM 为用户提供了一个集设计、分析、制造功能于一体的系统，用户只需动动手指就能使用。此外，其他一些日常的单调的（但是重要的）任务，比如说材料清单的准备、成本计算、生产调度等等也许可以通过使用同一电脑网络自动执行。使用 CAD/CAM 的另一个主要的优点是减少了从概念到设计再到制造的研制周期。产品开发的成本也能显著降低，因为像有限元法这样的分析可以在非常短的时间内达到最优化设计。

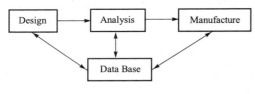

图 24-1 交互式计算机图表环境

CAD/CAM 及分析最适于用在交互式计算机图表环境中来解决如图 24-1 所示的工程问题。工件的几何形状和拓扑结构，以及其他的参数如材料性能等都存储在公共的数据库里，任何一个功能都能调用该数据库。像有限元法、上限元法和板坯分析法等分析都在初始的产品几何形状上进行。

通常用彩色显示出来的结果既可以自动也可以通过用户的介入来改变设计参数，以满足各种设计标准。这个过程可以不断地重复下去，直到获得最优的解决方案为止。然后，就能自动地或通过用户介入来执行制造和其他相关的功能。

CAD/CAM 系统

任何 CAD/CAM 系统都包括硬件——如中央处理器（CPU）、磁盘存储器、显示器、写字板等，以及软件——它是整个系统的大脑。有时当硬件与适当的软件安装在一起时，则很难清楚区分硬件和软件。这种情况通常被称为"固件"。

典型的 CAD/CAM 系统配置示于图 24-2 中。系统的核心是中央处理器，它可以协调 CAD/CAM 系统中的各项功能。用户通过工作站与系统交互作用，工作站由一个给用户提供可视化输出的图像显示器（或两个，如图）所构成。用户可以通过键盘或写字板（带有菜单）与系统交流。双屏工作站有很多优点。它可用于从两个或多个角度来观察正在设计的产品。可以通过放大命令在一个屏幕上放大产品的某一特征以供仔细观察，而另一个屏幕则用来观察产品的整体。在任何一个屏幕上都可以添加文本。其他供应商提供的系统带有图形显示器和用于文本的字符显示器。

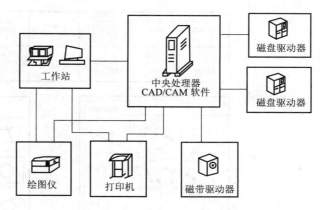

图 24-2 典型的 CAD/CAM 系统配置

工作站是智能的，因为它们可以在没有与 CPU 联系的情况下进行一些有用的局部操作，如放大、旋转等。目前的趋势是增强工作站的能力，从而解放 CPU 去执行其他更艰巨的任务。许多 CAD/CAM 系统有一个来自写字板或数字转换板的命令"菜单"可选择。菜单可以根据应用而变化。有些 CAD/CAM 系统使用按钮式菜单来选择某些命令。CAD/CAM 系统中使用的典型菜单如图 24-3 所示。

（三）阅读材料（13）参考译文

CAD/CAM 定义

CAD/CAM 是指计算机辅助设计和计算机辅助制造。它是一种运用数字计算机在设计和生产过程中执行某些功能的技术。这种技术正朝着更为综合的设计和制造的方向发展，而设计和制造在过去一直被看做是生产中两个有区别的、独立的过程。从根本上说，CAD/CAM 会为将来的计算机集成化工厂提供技术基础。

CAD 可以被定义为运用计算机系统对设计进行创造、修改、分析或优化。计算机系统由硬件和软件两部分组成，用来执行特定公司用户所要求的专门的设计功能。CAD 硬件一般包括电脑、一个或多个图形显示终端、键盘和其他外围设备。CAD 软件包括在系统上执行电脑绘图的程序以及使用户公司的工程功能简化的应用程序。这些应用程序的例子包括部件的应力应变分析、进程的动态响应、传热计算，以及数控零件编程。不同的用户公司的应用程序各不相同，因为其生产线、制造工艺和客户市场不一样。这些因素导致了对 CAD 系统的不同要求。

CAM 可以被定义为运用计算机系统并结合工厂里的生产资源直接或间接地对工厂的生产过程进行计划、管理和控制。如定义所示，计算机辅助制造的应用分为两大类：

1. 计算机监测与控制。这些是直接应用。为了监测或控制生产过程，将计算机与生产过程直接相连。

2. 制造支持应用。这些是间接应用，计算机用于支持工厂的生产操作，但是计算机与

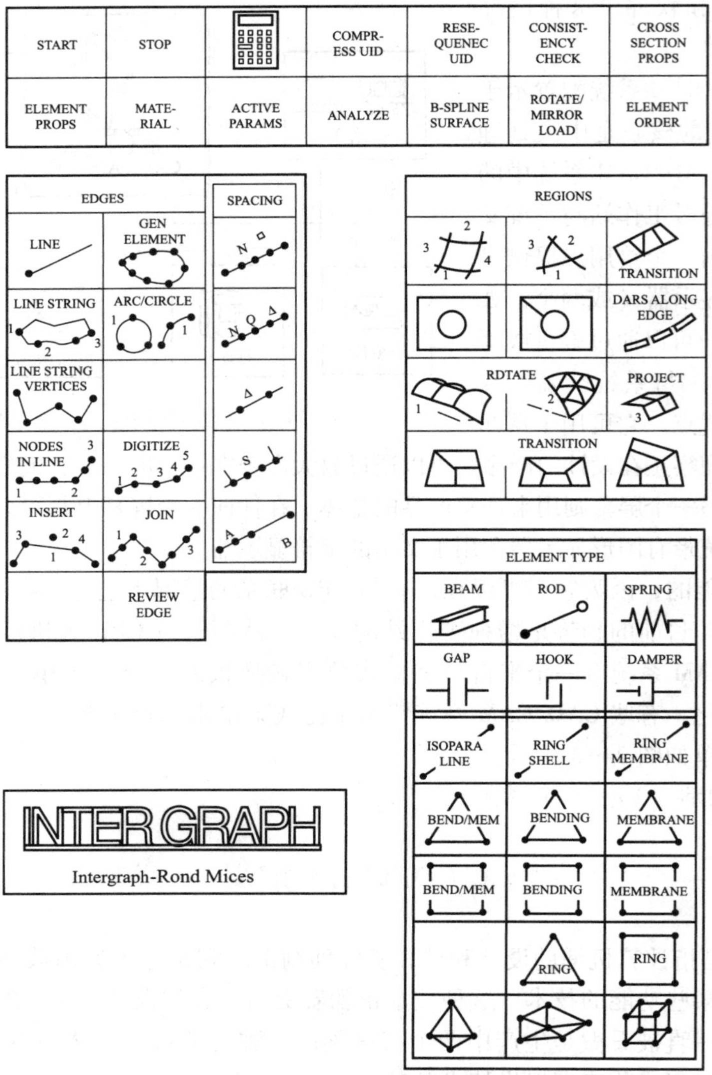

图 24-3 典型的 CAD/CAM 菜单（经 Intergraph 公司许可）

生产过程并没有直接联系。

这两者的区别对于了解计算机辅助制造是基本的。对这两种简单定义进行详细阐述是合适的。

计算机监测与控制可分为监测应用和控制应用。计算机工艺监测是计算机直接地介入生产过程，以监测加工过程、相关设备和收集加工数据为目的。计算机不是用来直接控制操作。对生产过程的控制仍然掌握在操作员手中，而操作员则以计算机所编译的信息作为指导。

计算机过程控制比监测更进一步，因为它不仅观测过程，而且还在观测的基础上控制过程。监测和控制两者之间的区别如图 24-4 所示。对于计算机监测，过程与计算机之间的数据流只是单向的，是从过程到计算机。而在计算机控制中，计算机接口可以进行双向数据流动。信号从过程传输到计算机，正如计算机监测中那样。此外，根据软件中包含的控制算法，计算机直接对生产过程发出命令信号。

图 24-4　计算机监测与计算机控制的对比
（a）计算机监测　（b）计算机控制

除了运用计算机对生产过程直接进行监控外，计算机辅助制造还间接地对工厂的生产过程起支撑作用。在这些应用中，计算机没有直接与生产过程相连。相反，计算机被离线用来提供方案、调度、预测、指示和信息，这样可以更有效地管理公司的生产资源。计算机与过程的关系形式如图 24-5 所示。虚线用来表明通信和控制链接是一种与人的离线连接，通常需要完善接口。用于制造支持的一些 CAM 例子有：

图 24-5　用于制造支持的 CAM

通过计算机进行数控零件编程。给自动化机床编写控制程序。

计算机自动的工艺设计。计算机准备好加工某一特定产品或部件所需要的操作工序列表。

计算机生成的工作标准。计算机为某一特定的生产操作确定时间标准。

生产调度。为满足生产需要，计算机确定一个适当的调度计划。

物资需求计划。计算机用来确定何时订购原材料和部件，以及需要订购多少来满足生产进度。

车间控制。在这个 CAM 的应用中，从工厂里采集数据来确定不同的生产车间的订单的进展。

在所有这些例子中，人的作用是输入计算机程序或者对计算机的输出结果进行解译，以完成所需的步骤。

Chapter 8

Quotation and Contract for Mold and Die
(模具的报价与合同)

Lesson 25

Introduction of Quotation for Mold

General Method of Quotation for Mold

Most of the molds that the professional factory of mold produce are single ones rarely with the same products, the same specification, and the same type of the mold. As for the mold, there is no unified pricing standard, so the same mold will have great differences in quotation at different areas or different enterprises, and some are even several times or more. A systematic material and scientific and reasonable calculation method haven't been formed in some mold factories.

Ways Generally Adopted in the Quotation for Mold

Analytical Method of the Historical Materials

When carrying on a newly mold, some professional product factories or professional mold factories usually tend to refer to the historic material, the price over the years, and then revise it a little and finally, the revised price is the quotation to the customers. Over two decades ago, some television molds even offered the price on the basis of the screen display. Take the black and white television for example, the shell mold of it with 14 in (36 cm) was 140 million RMB, and that of the 17 in (43 cm) cost 170 million RMB and so on. It is obviously that such quotation is rather raw. Some factories manufacturing fans also offered the price consulting the quotation of the previous years, and this way of quotation was also accepted by the customers at that time. Nowadays, a few enterprises still adopt such method to offer the mold.

Comparative Method of the Same Product

When developing the molds such as the leaf of the fans, several factories revised the current manufacturing price of the same type of mold in China and then offer the price to the outside. As for the comparative quotation of the same product, the manufacturing factories have relatively assurances, and the customers who often have business contact with the factories are relatively easier to accept the price. Some factories consult the quotation of American-European countries, Japanese or Hong Kong and Taiwan area in China and then offer the price to the outside with a discount coefficient separately. When the discount coefficient is suitable and also meets the requirement of quality of product as well as the delivery date, this way of quotation is much easier accepted by the customers. Certainly, in such aspects as manufacturing technology, enterprises must have suitable assurance.

Experiential Estimating Method of Material Rate of Exchange

According to the statistic of a number of data made by mold manufacturing, both the material cost and labor cost have certain proportional range, which are also showed abroad. However, we can't refer to it completely because of the big differences between material cost and labor cost at home and abroad. In quotation for mold, the method of experiential estimating way of material rate of exchange is also adopted. For example, some factories adopt the material price as the basis of rate of exchange, some range from 5 to 10 times of the material price, under the price complemented by other referential data. Of course, this estimating way is low level and should have differences according to the different mold.

Method of Fixing Price according to the Customs' Psychological Price

It's necessary for the customer to conduct an overall estimation of technology, marketing need and economic feasibility when developing a new product. That is to say, an initial scientific analysis is needed when developing a product and by this way, we have an expected price or psychological price for the mold. The manufacturer calculates regressively after knowing the customer's psychological price and then offer the price to him to seek the further way of negotiation.

Quotation method of cost prediction

The historic, analogical and experiential quotation is more and more restricted due to the fierce market competition and is hard to be accepted by the customers who attach importance to the management of enterprise. Furthermore, many enterprises ask for the detailed quotation quickly, or otherwise, it is difficult for the manufacturer to get order from the customers in the fierce market competition. With the development of science and technology together with the improvement of the managerial level of enterprise, the accurate quotation of mold and computer-assisted quotation will be valued and become an indispensable way for the mold manufacturer. We can also combine the command-centered quotation with the competition-centered quotation on the basis of the cost-centered quotation to meet the higher requirements of the customers and the market.

General Description of Injection Mold

The quotation for mold is an important economic factor to close a deal between the mold manufacturer and the customers. When the object of customers is an international aggregation enterprise, a foreign-funded enterprise or the leading one of the line, they will inspect, estimate and affirm the manufacturer according to the quality authentication system of ISO9000. The purchasing business can't come to a deal only by one department such as purchasing department or one person, it should be verified and estimated by the technical department, purchasing department and quality department. A very important big project should be submitted to the manager in chief to be examined and approved. Some international aggregation companies also formulate that a professional confirmation organization should examine a set of suppliers provided by the local companies. Only after the approval, can the local companies purchase from the qualified suppliers in the forward selection list. As for the mold manufacturer, every norms of achievement of delivery time, quality, cost and after-service will have a direct impact on the result that whether the suppliers can be listed in the supplier name list. At present, because customers have a higher demand to the quotation for mold, each company design a special quotation submitted to the supplier for filling in. Some quotations are divided into detailed rules for the customers to compare the quotation between different suppliers or even the same item and inquiry the supplier on the quality.

Mold manufacturers can get the following inspirations after the analysis of the above table list:

1) Quotation is the comprehensive reflection of technology, economy, efficiency and management of an enterprise, so the suppliers should put emphasis on the everyday statistic data and provide the advanced and quick mold needed by the users with the scientific management.

2) Aimed to the demands of the users, we should select material reasonably and apply the material economically to manufacture the high quality and low cost mold which meet the requirement of the customers.

3) Aimed to the demands of the users, we should work out manufacturing craft and select working equipment, and arrange labor together with technology reasonably to manufacture the mold which meet the requirements of the customers.

4) The process of quotation is an intense struggle for the customers between numerous suppliers who attack each other with all kinds of possible means but can't see each other in the commercial competition. It is obviously that it is a very important management work that can not be neglected for the enterprise to establish a quotation for mold system to meet the modern enterprise competition.

Quotation of Plastics Mold

In order to help you to master the quotation of mold, we give an example: a quotation of plastics mold of ××× molds & plastics products factory is shown in Tab. 25-1.

Tab. 25-1 Quotation of plastics mold

×××Mold & Plastics Products Factory					
Quotation of Plastics Mold					
Client:		Type:	Code:	Management No.:	
Part Description:				Remarks:	
Plastic Material		Plastic Weight		Surface Finish	
Cavity:	Kit Qty:	Mold Structure:	Steel for Cavity/Core:		Lead Time:
Detail Calculation					
(1) Material Cost					
Component		Type/Steel	Quantity	Specification/Size	Cost
Steel & Electrode	Cavity				HK $
	Core				HK $
	Insert				HK $
	Slider				HK $
	Copper				HK $
Mold Base		☐Edge Gate/ ☐Pin-Point Gate			HK $
		Hot Runner: With☐ Without☐			HK $
		Hot Sprue			HK $

continued

Hardware/Fitting						HK $
Remarks	Wedge		Slider	Other	Total	HK $
(2) Machine Cost				(3) Engineering Cost		
Item	Estimated Time of Machine	Price Per Hour	Cost	Item		Cost
CNC		HK $	HK $	Design Cost		HK $
Lathe		HK $	HK $	Project Following		HK $
EDM		HK $	HK $	Trial Shooting/3Times		HK $
Milling		HK $	HK $	Transportation		HK $
Grinding		HK $	HK $	Other		HK $
Polishing		HK $	HK $	Managing Cost		HK $
Wire Cutting		HK $	HK $	Total		HK $
Heat Treatment		HK $	HK $	Total Cost:		HK $
Texture		HK $	HK $	Mold Quoted:		HK $
Engraving Cost		HK $	HK $			
Modification Cost		Total:	HK $	Material Cost %; Engineering Cost %; Gross Profit %		
Prepared By/Date		Check/Date		Approve		

Words and Expressions

aggregation [ˌæɡrɪˈɡeɪʃən]　　　　　n.　集合；聚合；集合体
calculation [ˌkælkjuːˈleɪʃən]　　　　n.　计算；考虑
comparative [kəmˈpærətɪv]　　　　adj.　比较的；相当的
professional [prəˈfeʃənl]　　　　　adj.　专业的，职业的
quotation [kwəʊˈteɪʃən]　　　　　n.　引用语；价格；报价单；行情表
raw [rɔː]　　　　　　　　　　　　adj.　未加工的，处于自然状态的；生疏的
revise [rɪˈvaɪz]　　　　　　　　　vt.　修订，校订，修正，修改
supplier [səˈplaɪə]　　　　　　　　n.　供应者，补充者，厂商，供给者

1. Most of the molds that the professional factory of mold produce are single ones rarely with the same products, the same specification, and the same type of the mold.

译文：模具专业厂生产的模具多数为单件生产，同产品、同规格、同类型的模具几乎很少重复。

2. As for the mold, there is no unified pricing standard, so the same mold will have great

Chapter 8　Quotation and Contract for Mold and Die

differences in quotation at different areas or different enterprises, and some are even several times or more.

译文：对于模具，国内还没有统一的定价标准，所以同一个产品在不同的地区、不同的企业报价时往往会出现很大的差异，有的甚至差几倍或更多。

解析：**as for sb/sth**：至于/对于某人或某物。

3. When carrying on a newly mold, some professional product factories or professional mold factories usually tend to refer to the historic material, the price over the years, and then revise it a little and finally, the revised price is the quotation to the customers.

译文：有的专业产品厂或专业模具厂在承接一个新品模具时，往往会以历史的资料即历年的价格作为参考，并稍作修正，并最终以此来作为对客户的报价。

解析：tend to：倾向，趋向。refer to sb/sth：向某物或某人查询信息。

4. Over two decades ago, some television molds even offered the price on the basis of the screen display.

译文：20多年前，一些电视机模具甚至以显示屏的尺寸作为报价的依据。

解析：on the basis of：以……为中心、根据、基础。

5. When developing the molds such as the leaf of the fans, several factories revised the current manufacturing price of the same type of mold and then offer the price to the outside.

译文：在研发电扇风叶等模具时，不少工厂是以同类型模具的现行制造报价稍作修正后对外报价的。

6. According to the statistic of a number of data made by mold manufacturing, both the material cost and labor cost have certain proportional range, which are also showed abroad.

译文：据模具制造业大量资料的统计，材料费和人工费均有一定的比例范围，国外对此也有资料统计，但由于国内的材料费和人工费与国外有很大差异，所以不能完全参考。

7. It's necessary for the customer to conduct an overall estimation of technology, marketing need and economic feasibility when developing a new product.

译文：当客户在开发一个新产品时，对这个新产品需作出全面的技术评估、市场需求评估及经济可行性评估。

解析：it is+*adj*+for sb to do sth：对于某人来说做某事很困难。介词 for 后面的宾语为不定式动作的执行者，即逻辑主语。

8. The historical, analogical and experiential quotation is more and more restricted due to the fierce market competition and is hard to be accepted by the customers who attach importance to the management of enterprise.

译文：由于市场竞争的激烈，那种历史的、比拟的、经验的报价方法已越来越受到限制，也难以让注重企业管理的客户所接受。

解析：be due to：由于，应归于。attach importance to：认为……重要。

9. A very important big project should be submitted to the manager in chief to be examined and approved.

译文：至关重要的大项目应该提交给主管予以审核和批准。

解析：submit to：向……提交……（以供考虑、裁决等）。

Exercise

1. As for the mold, there is no _____ pricing standard, so the same mold will have great _____ in quotation at different areas or different enterprises, and some are even several times or more.
2. A systematic material and _____ and _____ calculation method haven't been _____ in some mold factories.
3. It's necessary for the customer to _____ an overall estimation of technology, marketing need and _____ feasibility when developing a new product.

Reading Materials (14)

Section A Quotation Strategies and Terms of Payment

The quotation and payment of mold is the continuity and result of the estimating mold. Estimate and quotation is just the first step. The final aim is to form balanced price after settlement of making and using of mold. During the process, people usually hope that estimated price is equal to formed price and settlement price. While in practical operation, the three prices are not completely equal. There is fluctuate error value between them. That's the problem we will discuss. After estimating the price, people need to coordinate the price and prepare for making machining mold contact. Through reiterative negotiation, the contact parties confirm the deliberative mold price, make contact and then manufacture molds.

1. Estimated Price, Quoted Price and Formed Price

Estimated price can not be directly used as quoted price. Generally speaking, people must coordinate the estimated price according to the synthetical analysis of market, customers' psychology and competitive adversary. The first quoted price must be increased by 10% to 30% based on the estimated price. After bargaining, sellers may decrease quoted price due to situation. When the negotiated price is low to 90% of the estimated price, sellers must reestimate the price because sellers must insure their margin on making contact with buyers. The price of mold which is confirmed by two parties and signed on the contact may be higher or lower than the estimated price. When the negotiated price is lower than the cost price, the sellers may propose to modify require, condition and plan so as to reduce the mold cost. After adjusting accounts, sellers make contact on modificative price. To point out, mold is a kind of specialized product of high tech. Sellers should not sell mold at low price or even loss price to cater for customers. On the contrary, sellers should pay much attention on the quality, precision, longevity of the mold. Otherwise, it may misdirect sellers to excessively pursue low price, which can not insure the quality, precision, longevity of the mold. Cheapness is not tolerated by mold industry. But, when the manufacture of mold and production of goods are belong to same accounting company or related economic unit, the quoted price of mold is based on the cost price. Estimated price only includes cost price. Margin and other cost are compensated through product margin and additional cost of mold. On this condition, the quoted price which can not be considered as real price of mold is used as forepart exploitation cost. Once the product exploited successfully and margin brought, seller should draw out additional cost of mold and return it to the mold-producing company. Cost and margin form new price of mold. New price may be higher than the former price, even several times than it. Of course, the return rate is nothing.

2. Different Price in Different Regions and Times

In different companies, regions and countries, estimated price and formed price of mold vary; its connotation changes in different condition. That is to say, there exists region difference and time difference. Why do price differences appear? On the one hand, the manufacture conditions in companies, regions and countries are diverse. Different equipments, technology, personnel conception and consumption level influence the estimate of product cost and margin so as to produce different mold price. Generally speaking, developed regions or standard large-scale companies with high-tech pursue producing high-quality and high-price products. While, in low consumption level or low-tech regions, some secondary companies with few equipment have few enquires to the price of mold. On the other hand, there is time difference between mold price, that is time efficiency difference. Different time enquires produce different mold price. The time efficiency consists of two connotations; one is that price is different at different time, and the other is that price is affected by manufacture cycle.

3. Filling in the Mold Quotation

After estimating mold price, sellers quote price to the public. The quotation consists of quoted price, cycle, required longevity die life, technology requires and conditions, terms of payment and guarantee deadline. Quotation strategies directly affect the price, margin and exertion of producing technology and management level. Thereby, quotation strategies are important embodiment of the successful company management.

4. Terms of Payment

Payment is the final aim of mold design and manufacture. The final price is that subject to the settled price, that is settlement price, which is the actual mold price.

Terms of payment start from the design of mold and go with the whole manufacture process. It is over until the design comes into fulfillment. Sometimes, it operates a long time. All the quality problems in design process would finally turn into economic settlement, which actually evaluates the technology quality of design and manufacture.

Terms of payment is proposed from the quotation and operate with the design of mold when the contact is signed. Different terms of payment present difference between design and manufacture. Terms of payment vary in regions and companies. With the development of market economy, there is a set of criterion and convention.

(1) "Half-to-half" model payment. Buyers pay 50% down payment in advance. Remaining part must be settled up after check and accept.

This term of payment is popular in early mold companies. Its advantages and disadvantages are as follow:

① 50% down payment are not enough to pay mold manufacture cost. The manufacture companies need launch money into it. That is to say, 50% down payment can not operate with the whole payment mold. Therefore, for the mold-producing companies, there exists great investment venture.

② After checking the molds, buyers pay for the remaining money. Fare for guarantee of mold is irrelative to the settlement.

③ When settling the remaining money, there is liable to defaulting the settlement because molds are completed.

④ In case that molds failed only 50% of the down payment is returned.

(2) "Six-to-four settlement" model. Buyers pay 60% down payment to sellers when two parties make contact. And buyers pay the remaining to sellers after checking the molds. This term of payment is similar to the first model, but it is of benefit to mold-producing companies because of the added 10% down payment.

(3) "Three-four-three settlement" model. Buyers pay 30% down payment, and repay 40% down payment after checking the design and sellers' preparing for materials. The last 30% must be settled up in a week after checking and accepting the mold.

This term of payment is popular in modern mold companies. Its characters are the following:

① The 30% down payment is used as earnest money.

② According to the check, buyers check rate of progress and its reliability, then repay 40%. This way strengthens the supervision to producing mold.

③ Buyers pay the remaining 30% payment after checking, accepting and using molds. This term of payment operates simultaneously with the design and manufacture of molds.

④ In case that molds failed, mold-producing part must return all the down payment and extra compensation, which is always one or two times of the down payment.

(4) Way of drawing additional mold cost from product margin. When sellers designing and producing mold, buyers only invest small amount of money to insure the basic cost of producing mold or not invest at all. After molds are used, buyers start to produce products. Buyers return a part of margin as cost of molds to sellers due to the amount of products. This way systematically links two parties and margin, investment venture and use efficiency, technology, economy, quality and producing efficiency, which extremely present molds' value and venture. This way is the currently developed tendency. It advances at make good use of two parties' advantages and currency investment. But for the mold-producing party, the venture is greater, but the return rate is higher.

All different terms of payment have a common point, that is to systematically link mold technology and economy target, and produce common efficiency. Also, it realizes estimated price to

Chapter 8 Quotation and Contract for Mold and Die

quoted price, then to contact price and settlement price, which is the real mold price. To try best to joint mold price with international convention and produce high quality and efficiency molds is the final object of molds' design and manufacture.

Words and Expressions

account	n.	计算；账目
adversary	n.	敌手，对手
continuity	n.	连续性，连贯性
embodiment	n.	体现；具体化，
market	n.	市场；销路；行情
negotiate	v.	（与某人）商议，谈判，磋商；买卖；让渡（支票、债券等）
payment	n.	付款，支付，报酬；偿还；报应，惩罚
settlemen	n.	沉降；解决；结算
strategy	n.	策略

1. After estimating the price, people need to coordinate the price and prepare for making machining mold contact

译文：模具估价后，需要进行适当处理，为签订模具加工合作依据。

2. Generally speaking, people must coordinate the estimated price according to the synthetical analysis of market, customers' psychology and competitive adversary.

译文：一般说来，还要根据市场行情、客户心理、竞争对手等因素进行综合分析，对估价进行适当的整理。

3. The price of mold which is confirmed by two parties and signed on the contact may be higher or lower than the estimated price.

译文：经过双方认可且签订在合同上的模具价格有可能高于估价或低于估价。

4. When the negotiated price is lower than the cost price, the sellers may propose to modify require, condition and plan so as to reduce the mold cost.

译文：当商讨的模具价格低于模具的保本价格时，卖方可重新提出修改要求、条件、方案等，以降低模具成本。

5. The quotation consists of quoted price, cycle, required longevity die life, technology requires and conditions, terms of payment and guarantee deadline.

译文：报价单的主要内容有：模具报价，周期，要求达到的模次（寿命），对模具的技术要求与条件，付款方式以及保修期。

6. All the quality problems in design process would finally turn into economic settlement, which

actually evaluates the technology quality of design and manufacture.

译文：设计制造中的所有质量问题会最终转化到经济结算方面来，实际上，经济结算是对设计和制造的技术质量的评价。

7. Buyers pay the remaining 30% payment after checking, accepting and using molds. This term of payment operates simultaneously with the design and manufacture of molds.

译文：买方在模具验收、接受和使用模具后，支付30%的余款。这种方式与模具的设计和制造同步运行。

Section B Computerized Price Quoting System for Injection Mold Manufacture

Introduction

Users Interface

The CQSIM software package does not require any mold making experts for data input. Moreover, there is no need for the users to have on hand mold making experience. Instead, users are recommended to have a basic knowledge of plastic injection molds. They should know the mold building process from a given product or drawing. This is because the user must be able to choose a suitable process and materials for a mold. Without correct input, accurate results cannot be obtained. The terms used in the data input forms, like mold base, ejector pin, cooling line, pocket and machining methods are familiar to mold makers, even new learners. Therefore, they are easy to fill in by cost estimators who have the basic ideas of injection molds. The main menu of the CQSIM (Fig. 25-1) will be shown automatically to users when the programme is selected from the Microsoft Access environment. Users can choose a suitable operation from the menu for the desired tasks by pressing a button on it. For cost estimation, a user only needs about 10-30 min (a longer time is needed for very complexity products) to fill in the blanks of the forms as guided by the programme. The results can be obtained within a minute of calculation. It is therefore very user friendly.

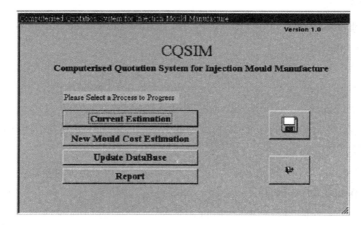

Fig. 25-1　The user interface of the CQSIM

User Input

There are six data input form input data for a mold. The main data that should be input into the system are:

(a) The contents of the products in the mold, e.g. client name, part number, part name, plastic material required, and their size.

(b) Mold base details, e.g. width, length, and height of individual plates.

(c) The general structure of a mold, e.g. ejector size and quantity, bosses and sleeve ejector (if any), and ribs and flat ejectors (if any).

(d) The part number (optional) and name of an insert.

(e) The machining details of the above insert, e.g. size, material, cooling line, subcontract, location of its pocket (hole to accommodate the insert), and machining methods to achieve the shape of the product.

Fig. 25-2 shows one of the six data input forms. Users can switch among these forms by pressing the buttons "Previous" or "Next". To start the calculation, press the "Finish" button. Users can also browse among different records in a form by the navigating buttons.

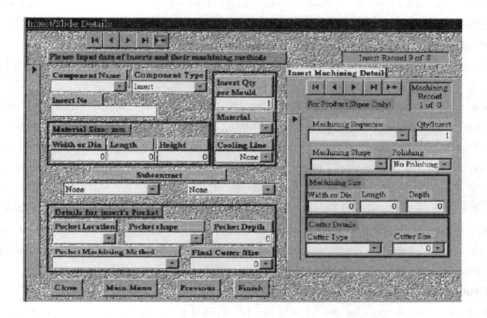

Fig. 25-2 A blank data input form for machining details

Data Tables

Four types of data tables are used to store different types of data:

(a) Input data:

(i) Product details, such as name of client, product size and polymer required.

(ii) Insert names, insert dimensions, and machining criterion.

(iii) General information such as size of mold base, gate type, and simple structures (e.g.

ejectors, ribs and bosses). Cost is calculated based on these data together with the price and technical data stated below.

(b) Resultant data:

(ⅰ) Materials cost, labor cost and machining cost.

(ⅱ) Total machining time for individual components and inserts.

(c) Price of standard components:

(ⅰ) Price for standard components such as different types of mold bases and ejectors.

(ⅱ) Price for different types of materials such as steel and copper.

Note: The price in these data tables should be updated by the users when their prices are changed.

(d) Technical data:

(ⅰ) Machining data, such as MRR and machining sequence.

(ⅱ) Standard time for dedicated activities, such as the time to compile a CAM program.

Machining Time Calculation

The CQSIM is developed independently from CAD-CAM systems. Therefore, the shape and volume for machining time calculation are not actual values but factorized values. The factors used in the CQSIM include Material Factor and Shape Factor. They are used in calculating the actual machining time from the "standard" machining time by means of MRR directly:

Actual machining time = "Standard" machining × Quantity × Shape factor × Material factor + Preparation time

Where preparation time includes set-up times, tear-down times and tool changing times, etc. The Shape Factor is used to compare the machining difficulties between individual shapes. Simple shapes like round, rectangular and drilling are assigned a value as "standard". Other shapes would be more difficult for machining (lower efficiency) and hence they have higher values. Material Factor is designed with reference to the hardness of the material for machining. The values for Material Factor can be assigned by referring to data given by manufacturers such as SKF or IDIA etc.

The factors in the CQSIM can be altered or added to when necessary by the users to suit their needs. Since the altered or newly created factors are done by the users themselves, they can grasp their meaning more easily to maintain the accuracy of the cost estimation.

Evaluation of the Prototype CQSIM

The quotation accuracy of the CQSIM can be divided into four parts. These are direct material costs, machining time for components, overheads, and profit. The overheads and profits are set by the users before the first time they use the CQSIM, by using the "Update Database" function of the CQSIM. Therefore, the accuracy of these two types of factors is affected by the set-up date from users. Thus, the accuracy of the CQSIM is based on direct material cost and machining time. Fig. 25-3 shows the result after a trial running of the CQSIM for analysis.

Accuracy of Estimated Direct Material Cost

The cost for materials is simply obtained by searching the database or by calculating the weight of the raw materials for mold components, so the accuracy is high (up to 98%). standard

Chapter 8 Quotation and Contract for Mold and Die

Insert Name	Description (Material Cost)	Time	Cost
Mould Base	Submarine Gate 3060 A100 B35 C125	0	$5,710
Angle Pin	Angle pin for Pin Hole Slide Dia 10 X 35	0	$5
Ejector Pin	Ejector Pin: Dia 6 X 130 mm	0	$175
Ejector Pin	Ejector Pin: Dia 3 X 130 mm	0	$40
Main Cav	150X210X100 Electrode $2041.05	187.1	$7,493
Etching	150X210X100	3	$575
Main Core	150X210X100	69.9	$6,653
Pin Hole Slide	60X75X25	10.9	$171
Rear Side Insert	36X50X35	6.9	$185
Machining Time for Submarine G	36X50X35	2.2	$28
Support Pillar	36X50X35	4.7	$243
(Boss Machining Time)	36X50X35 Electrode $86.75	13.6	$295
(Rib Machining Time)	36X50X35 Electrode $275.83	36.1	$814
(Ejector Machining Time)	36X50X35	5.4	$64
Mould Testing	36X50X35	12	$1,368
	Material Cost	0	$12,301
	Machining Cost	351.8	$11,438
	Component Cost (Material & Machining Cost)	0	$23,819
	Overhead Cost	0	$13,362
	Total Cost	0	$37,181
	Total Price	0	$41,313

Fig. 25-3 Accuracy of estimated direct material cost

components are supplied with current market price and counted in quantity required. Given the correct quantity and updated price, a more accurate cost can be obtained. The error for standard components are generated from the predefined quantity for angle pins, screws, etc. because the final choice may be slightly different to the programme. However, these components are generally cheap (a few dollars) and of small usage quantity. The error is therefore negligible when compared with the thousands of dollars total material cost. The 2% error comes mainly from the length of raw materials cut by suppliers. Generally, ironmongers will supply raw materials 5 – 10 mm longer than the required size.

The CQSIM treats it as 7.5 mm. For main cavity and main core insert of length 210 mm, the error calculated is only 1.2%. Then, larger inserts would yield smaller error, and vice verse. In most cases, the main inserts for molds are greater than 100 mm, so the total error for all materials would be very low. On the other hand, additional materials are also included in the calculation, e.g. cost for copper electrodes.

This will also increase the accuracy of the material calculation compared with empirical cost estimation. This is because electrodes, especially small electrodes, would be ignored or have their size underestimated in empirical cost estimation. The general practice in mold cost estimation is to estimate the cost of expensive components, e.g. mold base, main inserts, and main electrodes. A percentage charge for miscellaneous materials, e.g. electrodes for ribs, and shafts for bosses, would then be added. Doing it in this way, the error would sometimes be very large, and may be up to 30% or more. The errors are usually due to the size of the inserts, with this being overlooked. For example, the incremental size for some raw material may be 30 mm, so if the cost estimator treats it as 5 mm it would cause great error, and hence, their rough machining time estimation would be wrong. In some case, miscellaneous components would be ignored.

Accuracy of estimated machining time

The accuracy analysis of machining time should be compared with actual machining time of dedicated work-pieces. To have a precise analysis, work-studies should be carried out to examine the actual difference. Here, comparison is carried out empirically with some components of similar size and shape.

The machining time of the trial run for the components are shown in Table 25-2 The polishing for the main cavity is for both the cavity and the electrodes to machine its shape. Generally, it needs about 3 days (10 h per day) for each cavity, i.e., 60 h for two cavities. The electrodes for the 13 EDM processes need about 6 h to polish, so that the time required is reasonable because copper can be polished easily and a rough polish of the electrodes is acceptable. The CNC milling process is carried out for electrodes of the main cavity and the main core insert. Copper electrodes are extremely easy to machine, so the lead-time, 20.66 his much shorter than for the steel core, 16.19 h, and even the quantity of the former is much greater than that of the latter, a craftsman requires about 352 h (say 35 days) to finish the mold. Generally, adepts would quote it as 30 days. However, empirically, it is found that mold makers can seldom deliver the completed molds on time.

Table 25-2 The trial run result for machining time

Insert name	MC time (h)										
	Milling	Turning	Grinding	Drilling	Polishing	CNC milling	EDM	Wire cut	Labour	CAM	Total
Main cavity	39.26	0.69	5.19	7.57	65.66	20.66	25.62	0	73.94	1.25	187.1
Etching	0	0	0	0	0	0	0	0	3	0	3
Main core	1.93	0	1.08	7.57	10.55	16.19	0	23.96	16.97	2.25	69.92
Pin hole slide	4.37	0	1.78	1	0	0	0	0	10.95	0	10.95
Rear side insert	2.43	0	0.37	3.53	0	0	0	0	6.93	0	6.93
Submarine gate	0.54	0	0	1.51	0	0	0	0	2.15	0	2.15
Support pillar	0	2.51	0	1.92	0	0	0	0	4.68	0	4.68
Bosses	6.6	0.1	0.6	0	0.2	0	3.68	0	9.68	0	13.56
Ribs	15.93	0	2.09	0	0.7	0	8.16	0	27.27	0	36.13
Ejector	0	0	0	5.37	0	0	0	0	5.37	0	5.37
Mold test	0	0	0	0	0	0	0	0	12	0	12

Therefore, 32-33 days would be more appropriate. Thus, the error is about 9%. However, the estimation is not worse than manual estimation.

Words and Expressions

client	n.	顾客，客户；委托人
complexity	n.	复杂(性)；复杂的事物
dedicated	adj.	专注的；献身的
desired	adj.	渴望的，想得到的
empirically	adv.	以经验为主地
injection	n.	注射；注射剂
obtain	vt.	获得，得到

Notes

1. The CQSIM software package does not require any mold making experts for data input.

译文：CQSIM 软件包不需要任何模具制造专家来输入数据。

2. Instead, users are recommended to have a basic knowledge of plastic injection molds.

译文：相反，使用者只需要知道有关塑料注射模具方面的基本知识即可。

3. The main menu of the CQSIM (Fig. 25-1) will be shown automatically to users when the programme is selected from the Microsoft Access environment.

译文：当程序能够在微软数据库环境运行时，CQSIM 的主菜单（见图 25-1）将自动显示给使用者。

4. The accuracy analysis of machining time should be compared with actual machining time of dedicated work-pieces.

译文：对加工时间的准确分析应当与实际用于零件的机加工时间相比较。

课文导读及参考译文

（一）课文导读

模具报价是模具制造商和客户之间能否成交的一个重要经济条件。本课介绍了以下几种报价方法：历史资料分析法；相同产品比较法；材料比价经验估算法；客户心理价位定价法；成本预测报价法。同时也简要介绍了注射模具报价。

（二）课文参考译文

模具报价概述

模具报价的一般方法

模具专业厂生产的模具多数为单件生产，同产品、同规格、同类型的模具几乎很少重复。对于模具，国内还没有统一的定价标准，所以同一个产品在不同的地区、不同的企业报价时往往会出现很大的差异，有的甚至差几倍或更多。在一些模具工厂还没有形成系统的资料和科学合理的计算方法。

经常采用的模具报价方法

历史资料分析法

有的专业产品厂或专业模具厂在承接一个新品模具时，往往会以历史的资料即历年的价格作为参考，并稍作修正，并最终以此来作为对客户的报价。20多年前，一些电视机模具甚至以显示屏的尺寸作为报价的依据。以黑白电视机为例，14英寸（36厘米）电视机的机壳模具为14万人民币，17英寸（43厘米）电视机的机壳模具的费用是17万人民币，等等。这样的报价显然相当粗糙。一些生产风扇的工厂也参照以前的报价来报价，这种报价方法在当时也被顾客所接受。现在，仍有一些企业采用这种方法给模具报价。

相同产品比较法

在研发电扇风叶等模具时，不少工厂是对同类型模具的现行制造报价稍作修正后对外报价的。对于相同产品的比较报价，制造厂比较有把握，与工厂经常有业务往来的客户也比较容易接受。有的工厂参照欧美国家、日本或中国港台地区的报价，然后分别给以一个折扣系数而对外报价。当折扣系数合适，而且也满足产品质量和交货日期的要求时，这种报价方法更容易被客户接受。当然，在制造技术等方面，企业一定要有相当的把握。

材料兑换率的经验估价法

据模具制造业大量资料的统计，材料费和人工费均有一定的比例范围，国外对此也有资料统计。但由于国内的材料费和人工费与国外有很大差异，所以不能完全参考。在模具报价中，也采用材料兑换率的经验估价法。例如，一些工厂采用材料价格作为兑换率的基础，有的控制以5~10倍的材料价，辅之以其他参考资料作为模具的报价。当然，这种估价法是低层次的，而且应该根据不同的模具而异。

根据客户心理价位定价法

当客户在开发一个新产品时，对这个新产品需作出全面的技术评估、市场需求评估及经济可行性评估。也就是说，在开发产品时需要有一个初步的科学分析，通过这种方法，我们对模具就有了一个期望价位或心理价位。当制造商知晓客户的心理价位后，再作逆向计算，然后对客户作一个报价，以寻求进一步谈判的途径。

成本预测报价法

由于市场竞争的激烈，那种历史的、比拟的、经验的报价方法已越来越受到限制，也难以让注重企业管理的客户所接受。而且，不少企业要求迅速作出详细的报价，否则，在激烈的市场竞争中，制造商就难以从客户手中拿到订单。随着科学技术的发展以及企业管理水平的提高，精确的模具报价和计算机辅助报价会逐渐受到重视并成为模具制造商不可缺少的一种手段。在以成本为中心的报价基础上，同时也可以结合以需求为中心的报价和以竞争为中心的报价，以满足客户和市场的更高要求。

注射模具报价概述

模具报价是模具制造商和客户之间能否成交的一个重要经济因素。当客户对象为国际集团企业、外资企业、居行业领袖地位的企业，他们会按ISO9000质量认证体系的要求对模具制造商进行考察、评估和确认。采购业务也不是由一个部门如采购部或一个人说了算，而应该由技术部门、采购部门和质量部门来审核与评估。至关重要的大项目应该提交给高级主管予以审核和批准。有的国际集团公司还规定专业的确认机构应该审核当地公司提供的一批供应商。只有同意之后，当地公司才能从预选名单中审定合格的供应商处采购。对于模具制造商而言，交货期、质量、成本、售后服务每一项指标的绩效，均会直接影响到本企业能否被列入供应商名单中。现在，由于客户对模具报价有了更高的要求，因此各公司设计了交给供应商填写的专门报价单。有的报价单栏目分得很细，便于客户对不同供应商之间的报价或者甚至对同一项目进行比较，并向供应商进行咨询。

分析上述表单，模具制造商可得到以下启示：

（1）报价是企业技术、经济、效率和管理的综合反映，因此供应商必须重视日常统计资料，运用科学的管理提供用户所需的先进、快速的模具。

（2）针对客户的要求，制造商必须合理地选材，经济地用材，制造出符合客户要求的质优价廉的模具。

（3）针对客户的要求，制造商必须合理地制定制造工艺，选用加工设备，配置人力和技术，制造出符合客户要求的模具。

（4）报价的过程是众多供应商之间激烈的客户之争，供应商在商业竞争中以各种可能的方式相互进攻，但又互不照面。显然，建立符合现代企业竞争需要的模具报价系统，对于企业而言是一项不可忽视的、非常重要的管理工作。

塑料模的报价

为了帮助读者掌握模具的报价，我们给出了一个实例：表格25-1所示为×××模具与塑料制品厂的塑料模报价单。

表25-1 塑料模报价单

×××模具与塑料制品厂					
塑料模报价单					
客户：		类型：	代码：		管理号：
零件描述：				备注：	
塑料材料			塑料重量	表面光洁度	
型腔：		整套质量：	模具结构：	型腔/模芯用钢：	交付时间：
详细计算					
（1）材料成本					
组件		类型/钢	数量	规格/尺寸	成本
钢和电极	型腔				HK $
	模芯				HK $
	镶块				HK $
	滑块				HK $
	铜				HK $
模架		□侧浇口/ □点浇口			HK $
		热流道：有□ 无□			HK $
		热主流道			HK $
硬件/装配					HK $
备注	楔块		滑块	其他	合计 HK $

续表

（2）加工成本				（3）工程成本	
项目	估计的加工时间	每小时价格	成本	项目	成本
CNC		HK $	HK $	设计成本	HK $
车削		HK $	HK $	后续工程	HK $
EDM		HK $	HK $	试模/3 次	HK $
铣削		HK $	HK $	运输	HK $
磨削		HK $	HK $	其他	HK $
抛光		HK $	HK $	管理成本	HK $
线切割		HK $	HK $	合计	HK $
热处理		HK $	HK $	总成本：	HK $
纹理		HK $	HK $	模具报价：	HK $
刻模成本		HK $	HK $		
修模成本		合计：	HK $	材料成本 % 工程成本 % 总利润 %	
制表/日期		校核/日期		批准	

（三）阅读材料（14）参考译文

A 模具报价的策略和付款方式

模具的报价与结算是模具估价后的延续和结果。模具的估价和报价只是第一步。最终目的是在模具制造和使用完成后形成均衡价格。在这个过程中，人们通常希望模具估价等于形成价格和结算价格。而在实际操作中，这三个价格并不完全相等。三种价格之间有波动的误差值。这就是以下要讨论的问题。模具估价后，需要进行适当处理，为签订模具加工合同作依据。经过反复谈判，合同双方确定协商好的模具价格，签订合同，然后制造模具。

1. 模具的估价、报价与形成价

模具的估价不能直接作为报价。一般来说，还要根据市场行情、客户心理、竞争对手等因素进行综合分析，对估价进行适当的整理。第一次报价必须在估价的基础上增加10%~30%。经过讨价还价，卖方可根据实际情况降低报价。但是，当模具的商讨价低于估价的90%时，卖方需要重新估价，因为卖方必须保证与买方签订合同时的利润空间。经过双方认可且签订在合同上的模具价格有可能高于估价或低于估价。当商讨的模具价格低于模具的保本价格时，卖方可重新提出修改要求、条件、方案等，以降低模具成本。调整完价格计算后，卖方按修改后的价格签订合同。应当指出，模具是一种技术含量高的专业产品。卖方不应该以低价，甚至是亏本价去迎合客户。相反地，卖方应该更多关注模具的质量、精度、寿命等。否则，可能会误导卖方过分追求低价，这样就不能保证模具的质量、精度和寿命。廉价不是模具行业之所为。但是，当模具的制造与产品的生产属于同一会计公司或相关的经济单位时，模具的报价应以成本价为基础。估价仅包括成本价。利润和其他成本通过产品利润和模具的追加成本来补偿。在这种情况下，报价是作为前期开发成本，而不能被认为是模具的真实价格。一旦产品开发成功，获得利润，卖方应该提取模具的追加成本并返还给模具制造公司。成本和利润形成了新的模具价格。新的价格可能高于以前的价格，甚至会高出好几倍。当然，收益率为零。

2. 不同地区与时间的不同价格

在不同的公司、地区和国家里，模具的估价和形成价是不同的，其内涵在不同条件下会改变。也就是说，存在地区差异和时间差异。为什么会出现价格差异呢？一方面，各企业、各地区和国家的模具制造条件不一样。不同的设备、技术、人员观念和消费水平会影响产品成本和利润的估算，从而产生了不同的模具价格。总的来说，发达地区或者具有高技术的规范的大型公司追求的是生产优质高价的产品。而在消费水平低或者技术水平低的地区，一些只有少量设备的二流公司对模具价格没有什么要求。另一方面，模具价格之间还存在时间差异，即时间效率的差异。不同的时间要求会得到不同的模具价格。时间效率包含两个内涵：一个是不同时间的价格会不同；另一个是制造周期影响价格。

3. 填写模具报价单

模具价格估算完以后，卖方对外报价。报价单的主要内容有：模具报价，周期，要求达到的模次（寿命），对模具的技术要求与条件，付款方式以及保修期。报价策略直接影响价格、利润，以及生产技术和管理水平的发挥。因此，报价策略是成功企业管理的重要体现。

4. 付款方式

付款是模具设计和制造的最终目的。最终价格是已结算的价格，即结算价格，这是实际的模具价格。

付款方式从模具设计开始，贯穿着整个制造过程。直到设计完成以后付款方式才终结。有时，付款需要长时间的操作。设计过程中的所有质量问题会最终转化到经济结算方面来，实际上，经济结算是对设计和制造的技术质量的评价。

付款方式从报价时提出，签订合同后与模具设计同步运行。不同的付款方式体现了设计与制造之间的不同。结算方式在各地区、各企业均有不同。随着市场经济的发展，形成了一套标准和惯例。

（1）"五-五"式付款。卖方预先支付模具款的50%。余额必须在检验和接收以后结清。这种付款方式在早期的模具企业中比较普遍。其优、缺点如下：

① 50%预付款不足以支付模具制造成本。制造企业需要向其中投入资金。也就是说，50%预付款不能与整个模具制造同步运行。因此，对模具制造企业来说存在很大的投资风险。

② 检验完模具以后，买方支付余款。模具保修的费用与结算无关。

③ 在结算余款时，由于模具已经完工，易产生结算拖欠现象。

④ 万一模具失败，仅退回50%的预付款。

（2）"六-四结算"方式。双方签订合同时，买方向卖方预先支付模具款的60%。模具检验合格以后，买方向卖方结清余额。这种付款方式类似于第一种方式，但是这种方式对模具制造企业有利，因为预付款增加了10%。

（3）"三-四-三结算"方式。买方预付模具款的30%，审核完模具设计和卖方对模具材料的准备后，再支付40%的模具款。验收完模具后一周内必须结清最后的30%。

这种付款方式常见于现代模具企业中。其特点如下：

① 30%预付款作为订金。

② 根据检验，买方检查进度和可靠性，然后再支付40%的模具款。这种方式加强了对模具制造的监督。

③ 买方在检验、接受和使用模具后，支付30%的余款。这种付款方式与模具的设计和制造同时运行。

④ 万一模具失败，模具制造方必须返还全部预付款和额外的赔偿金，赔偿金一般是预付款的1倍或2倍。

（4）从产品利润中提取追加模具成本的方式。卖方设计和制造模具时，买方仅投入少量资金以保证制造模具的基本成本，或者根本不投入资金。模具使用后，买方开始生产产品。买方根据产品数量返还一部分利润给卖方作为模具成本。这种方式把合同双方与利润、投资风险和利用效率、技术、经济、质量和生产效率系统地连接起来，极大地体现了模具的价值和风险。这种方式是目前的发展趋势。它很好地利用了买卖双方的优势和货币投放。但是对于模具制造方而言风险较大，不过收益率也较高。

所有不同的付款方式都有一个共同点，那就是将模具技术和经济指标系统地结合起来，产生共同的效率。而且，付款方式实现了由估价到报价，然后到合同价和结算价，结算价是实际的模具价格。努力将模具价格与国际惯例接轨，生产优质高效的模具是模具设计和制造的最终目标。

B 用于注射模制造的计算机报价系统

概 述

用户界面

CQSIM 软件包不需要任何模具制造专家来输入数据。而且，用户不需要有现成的模具制造经验。相反，使用者只需要知道有关塑料注射模具方面的基本知识即可。他们应该从给定的产品或图纸知道模具的制作工艺。这因为用户必须能为模具选择合适的工艺和材料。没有正确的输入，就无法获得准确的结果。用于数据输入形式中的术语，如模架、推杆、冷却系统、模腔和加工方法已为模具制造者甚至初学者所熟知。因此，具有注射模基本知识的成本评估师能轻松填写报价系统。当程序能够在微软数据库环境运行时，CQSIM 的主菜单（图25-1）将自动显示给使用者。用户可以通过按键从菜单中为期望的任务选择合适的操作。对于成本估算，在程序的指导下用户只需花费大约10至30分钟（非常复杂的产品需要较长时间）来填写表格。一分钟之内就能得到计算结果。因此该系统对于用户来说极其方便快捷！

用户输入

对于模具，有6种数据输入形式。应该输入到系统中的主要数据是：

（a）模具产品的内容，如客户名称、工件数量、工件名称、要求的塑料材料和尺寸。

（b）模架的细节，如单个模板的宽度、长度和高度。

（c）模具的总体结构，如顶出装置的尺寸和数量、凸台和推管顶出装置（如果有），以及肋条和平面顶杆（如果有）。

（d）工件数量（可选择的）和嵌件名称。

（e）上述嵌件的加工细节，如尺寸、材料、冷却系统、局部收缩、空腔的位置（容纳嵌

Chapter 8　Quotation and Contract for Mold and Die

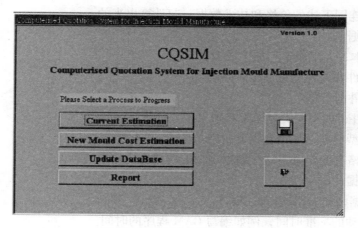

图 25-1　CQSIM 的用户界面

件的孔洞）、获得产品形状的加工方法。

图 25-2 所示为 6 种数据输入表格中的一种。用户通过点击按钮"上一个"或"下一步"可以在这些表格之间转换。点击"完成"键即可开始运算。用户也可通过导航键浏览表格中的不同记录。

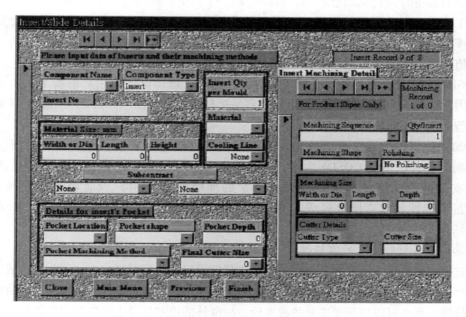

图 25-2　用于加工细节的数据输入空白表格

数据表

采用 4 种类型的数据表来储存不同类型的数据。

（a）输入数据：

（i）产品的细节，如客户名称、产品尺寸和要求的聚合物。

（ii）嵌件名称、嵌件尺寸和加工标准。

（iii）总体信息，如模架尺寸、浇口类型、简单结构（如顶出器、肋条和凸台）。成本是基于这些数据，连同下述的价格和技术数据一起计算出来的。

（b）结果数据：

（i）材料成本、人工成本和加工成本。

（ii）单个组件和嵌件的总加工时间。

（c）标准组件的价格：

（i）标准组件，如不同类型模架和顶出装置的价格。

（ii）不同类型材料，如钢和铜的价格。

注意：当价格有变化时，这些数据表中的价格应由用户更新。

（d）技术数据：

（i）加工数据，如 MRR 和加工顺序。

（ii）专用活动的标准时间，例如编写 CAM 程序的时间。

加工时间的计算

CQSIM 是从 CAD-CAM 系统独立开发而成的。因此，用于加工时间计算的形状和体积不是实际值，而是分解值。CQSIM 中使用的因素包括材料因素和形状因素两部分。直接通过 MRR，这些因素被用于从"标准"加工时间计算实际加工时间：

实际加工时间="标准"加工时间×数量×形状因素×材料因素+准备时间

式中的准备时间包括调整时间、拆卸时间、工具更换时间等。形状因素用来比较每个形状的加工难度。简单形状如圆形、矩形和钻孔被指定为"标准"值。其他形状会更难以加工（效率较低），因此具有较高的值。材料因素是参考被加工材料的硬度来设计的。材料因素的数值可以通过参考制造商如 SKF 或 IDIA 等提供的数据来确定。

CQSIM 中的因素可以在必要的时候由用户更改或增加以适应用户的需求。由于更改或新增的因素由用户自己完成，因此用户更容易把握自己的意图来保持成本估算的准确性。

原型 CQSIM 的估算

CQSIM 的报价准确性可以分为 4 个部分，即直接材料成本、组件加工时间、经费开支和利润。在首次使用 CQSIM 之前，通过使用 CQSIM 的"更新数据库"功能由用户设置经费开支和利润。因此，这两种类型因素的准确性受到用户设置数据的影响。所以，CQSIM 的准确性是以直接材料成本和加工时间为基础。图 25-3 所示为 CQSIM 分析试运行后的结果。

估算的直接材料成本的准确性

通过搜索数据库或计算模具组件材料的重量即可轻松得到材料成本，因此材料成本的准确率高（高达 98%）。标准件按目前的市场价供货，以所需数量计算。如果给定正确的数量和更新后的价格，那么可以得到更准确的成本价。标准件的误差是源于预先确定的斜销、螺钉等的数量。因为最终选择可能与程序中的稍有不同。然而，这些组件通常很便宜（几美元）而且使用量小。因此，与成千上万美元的总材料成本相比，这种误差是可以忽略的。2%的误差主要来自被供应商减少的原材料长度。五金商供应的原材料通常比要求的尺寸长 5~10 毫米。

CQSIM 把原料长度看作 7.5 毫米。对于长 210 毫米的主型腔和主型芯嵌件而言，计算的误差仅为 1.2%。此外，较大嵌件的误差较小，反之亦然。在多数情况下，模具的主要嵌件大于 100 毫米，因此所有材料的总误差非常小。另一方面，附加的材料也包括在计算之内，例如铜电极的成本。

Chapter 8 Quotation and Contract for Mold and Die

Insert Name	Description (Material Cost)	Time	Cost
Mould Base	Submarine Gate 3060 A100 B35 C125	0	$5,710
Angle Pin	Angle pin for Pin Hole Slide Dia 10 X 35	0	$5
Ejector Pin	Ejector Pin: Dia 6 X 130 mm	0	$175
Ejector Pin	Ejector Pin: Dia 3 X 130 mm	0	$40
Main Cav	150X210X100 Electrode $2041.05	187.1	$7,493
Etching	150X210X100	3	$575
Main Core	150X210X100	69.9	$6,653
Pin Hole Slide	60X75X25	10.9	$171
Rear Side Insert	36X50X35	6.9	$185
Machining Time for Submarine G	36X50X35	2.2	$28
Support Pillar	36X50X35	4.7	$243
(Boss Machining Time)	36X50X35 Electrode $86.75	13.6	$295
(Rib Machining Time)	36X50X35 Electrode $275.83	36.1	$814
(Ejector Machining Time)	36X50X35	5.4	$64
Mould Testing	36X50X35	12	$1,368
	Material Cost	0	$12,301
	Machining Cost	351.8	$11,438
	Component Cost (Material & Machining Cost)	0	$23,819
	Overhead Cost	0	$13,362
	Total Cost	0	$37,181
	Total Price	0	$41,313

嵌件名称	描述（材料成本）	时间	成本
模架	潜伏式浇口 3060 A100 B35 C125	0	$ 5 710
斜销	销孔滑块的斜销 直径 10×35	0	$ 5
推杆	推杆：直径 6×130 mm	0	$ 175
推杆	推杆：直径 3×130 mm	0	$ 40
主型腔	150×210×100 电极 $ 2 041.05	187.1	$ 7 493
刻模	150×210×100	3	$ 575
主型芯	150×210×100	69.9	$ 6 653
销孔滑块	60×75×25	10.9	$ 171
后侧嵌件	36×50×35	6.9	$ 185
潜伏式浇口加工时间	36×50×35	2.2	$ 28
支柱	36×50×35	4.7	$ 243
（凸台加工时间）	36×50×35 电极 $ 86.75	13.6	$ 295
（肋条加工时间）	36×50×35 电极 $ 275.83	36.1	$ 814
（顶出装置加工时间）	36×50×35	5.4	$ 64
试模	36×50×35	12	$ 1 368
	材料成本	0	$ 12 301
	加工成本	351.8	$ 11 438
	组件成本（材料与加工成本）	0	$ 23 819
	杂项开支	0	$ 13 362
	总成本	0	$ 37 181
	总价	0	$ 41 313

图 25-3 估算的直接材料成本的准确性

与以经验的成本估算相比，这也增加了材料计算的准确性。这是因为电极，特别是小电极，在经验的成本估算中被忽略，或者电极尺寸被低估。模具成本估算的通常做法是估算昂贵组件的成本，如模架、主要嵌件和主要电极。然后加上杂项材料，如用于肋条的电极、用于凸台的轴的费用。以这种方式进行估算有时会导致非常大的误差，误差可能达到 30%或者更大。误差通常是由于嵌件的尺寸被忽略而造成的。例如，某些原材料的增量尺寸可能是 30 毫米，因此如果成本评估师将其当作 5 毫米来处理，则会造成很大的误差。因此，原材料加工时间的粗略估计是错误的。在某些情况下，杂项组件可以忽略不计。

估算的加工时间的准确性

对加工时间的准确分析应当与专用工件的实际加工时间相比较。为了得到准确分析，应该进行工效研究来检查实际的差异。在这里，是与具有相似尺寸和形状的组件进行经验的比较。

组件试运行的加工时间列于表 25-2 中。主型腔的抛光是同时针对型腔和电极来加工其形状。每个型腔通常需要 3 天左右（每天 10 小时），即两个型腔需要 60 小时。用于 13 道 EDM（电火花加工）工艺的电极需要抛光大约 6 小时，因此要求的时间是合理的，因为铜容易抛光，而且电极的粗糙抛光是容许的。CNC 铣削工艺用于主型腔和主型芯嵌件的电极。铜电极极易加工，因此 20.66 小时的交付时间比钢质型芯的交付时间 16.19 小时要短得多，甚至前者的数量也远大于后者。人工需要大约 352 小时（即 35 天）来完成模具制造。熟练工通常会报 30 天。但是，根据经验，模具制造商很少能按时交付已完工的模具。

表 25-2 加工时间的试运行结果

嵌件名称	MC 时间（小时）										
	铣削	车削	磨削	钻孔	抛光	CNC 铣削	EDM	线切割	人工	CAM	合计
主型腔	39.26	0.69	5.19	7.57	65.66	20.66	25.62	0	73.94	1.25	187.1
刻模	0	0	0	0	0	0	0	0	3	0	3
主型芯	1.93	0	1.08	7.57	10.55	16.19	0	23.96	16.97	2.25	69.92
销孔滑块	4.37	0	1.78	1	0	0	0	0	10.95	0	10.95
后侧嵌件	2.43	0	0.37	3.53	0	0	0	0	6.93	0	6.93
潜伏式浇口	0.54	0	0	1.51	0	0	0	0	2.15	0	2.15
支柱	0	2.51	0	1.92	0	0	0	0	4.68	0	4.68
凸台	6.6	0.1	0.6	0	0.2	0	3.68	0	9.68	0	13.56
肋条	15.93	0	2.09	0	0.7	0	8.16	0	27.27	0	36.13
顶出器	0	0	0	5.37	0	0	0	0	5.37	0	5.37
试模	0	0	0	0	0	0	0	0	12	0	12

因此，32~33 天更加合适。误差大约为 9%。但是计算机估算比人工估算更好。

Lesson 26

Mold Making Contract

Purchaser:

Mold Maker:

This contract of mold making between the purchaser and mold maker, has been approved by the purchaser and mold maker. Terms and conditions stipulated below:

Ⅰ. Mold name, specification, steel for the cavities&cores, and price listed as below:

NAME	SPECIFICATION	STEEL FOR CAVITIES & CORES	PRICE	COMMENTS
		TOTAL		

Ⅱ. All the mold prices are FOB Guangzhou.

Ⅲ. The lead time of mold making: The lead time will be _____ days, from the mold maker confirms receipt of the 50% deposit to the mold maker sends the first articles.

Ⅳ. Mold design&manufacturing requirements:

a) The mold maker should design & make the molds according the samples or the drawings what provided by the purchaser, and conform to the purchaser's requirements.

b) The cooling water system should be appropriate so that the mold can be cooled enough.

c) The mold design should be approved by the purchaser before start making the mold.

d) If necessary, the purchaser should provide the specification and data of the injection machine which will work with the mold.

e) The parts should be fall down automatically when the mold run normally except some special parts.

Ⅴ. Payment terms: 50% of the total price should be prepaid as deposit before the projects of mold making starting, and the 50% final payment should be paid after the mold checked and approved. The mold maker should ship the mold to the purchaser after confirm the 50% final payment settled.

Ⅵ. Responsibility terms:

f) If the mold maker doesn't make the mold according the sample or drawing provided by the purchaser, the purchaser has the right to ask the mold maker repair or rework the mold.

g) The mold maker should accept to be forfeited money by the purchaser if the mold maker can't finish the mold in the appointed lead time. The 0.1% of the total mold price will be cut down as forfeiture of per day's delay.

h) If the purchaser need to change the mold or the part after the files which used to build the mold and mold design approval, the purchaser should pay for the additional changes, also the mold maker will free of the responsibility if the lead time delays. And the price of changes and the update lead time should be negotiated by the both parties.

Ⅶ. The official language of this contract is Chinese, and all the terms and interpretation should be according the Chinese version. This contract should be submitted to Chinese law. The other details that don't listed in this contract should be kindly negotiated by the both parties.

Purchaser:
Address:
Commissary:
(Signature) _____
Date: 2008-　-　　 (yyyy-mm-dd)
Mold Maker:
Address: Commissary:
(Signature) _____
Date: 2008-　-　　 (yyyy-mm-dd)

Words and Expressions

approved [əˈpruːvd]	adj.	经核准的，被认可的
commissary [ˈkɔmisəri]	n.	代表；物资供应所；委员
confirm [kənˈfəːm]	vt.	确定；批准；使巩固；使有效
	v.	确认
conform [kənˈfɔːm]	vt.	使一致；使遵守；使顺从
	vi.	符合；相似；适应环境
	adj.	一致的；顺从的
FOB = Free on Board		船上交货价格，离岸价格
forfeit [ˈfɔːfit]	n.	（因犯罪、过失、违约等而）丧失的东西，没收物；罚款
	vt.	没收；丧失
	adj.	丧失了的
forfeiture [ˈfɔːfitʃə]	n.	丧失；没收物；罚款

Chapter 8 Quotation and Contract for Mold and Die

mold maker		模具制造者，卖方
prepaid [ˈpriːˈpeid]	adj.	先付的，已支付的
purchaser [ˈpəːtʃəsə]	n.	买方，购买者
receipt [riˈsiːt]	n.	收条；收据；收到
	v.	收到
settled [ˈsetld]	adj.	固定的
signature [ˈsignitʃə]	n.	签名；署名；信号
specification [ˌspesifiˈkeiʃən]	n.	详述；规格；说明书；规范；型腔数
stipulate [ˈstipjuleit]	v.	规定，保证
submit [səbˈmit]	v.	（使）服从，（使）顺从
	vt.	提交，递交
update [ʌpˈdeit]	v.	使现代化；修正；校正；更新
	n.	现代化；更新
version [ˈvəːʃən]	n.	译文，译本，翻译

Notes

1. The mold maker should design & make the molds according to the samples or the drawings provided by the purchaser, and conform to the purchaser's requirements.

译文：模具制造商应根据买方提供的样品或图纸来设计和制造模具，并符合买方的要求。

解析：conform to：与某事物相符合或相一致。

2. Payment terms：50 % of the total price should be prepaid as deposit before the projects of mold making starting, and the 50% final payment should be paid after the mold checked and approved. The mold maker should ship the mold to the purchaser after confirm the 50% final payment settled.

译文：付款方式：模具开始制作前，预付模具总价的50%作为订金，模具验收合格后，付清其余50%，模具制造商确认收到所有付款后装运模具。

解析：deposit：定金，定钱。

3. The mold maker should accept to be forfeited money by the purchaser if the mold maker can't finish the mold in the appointed lead time. The 0.1% of the total mold price will be cut down as forfeiture of per day's delay.

译文：如模具制造商未能在指定交期内完成模具制造，则应接受买方的罚款。每延期一天将扣除模具总价的0.1%作为罚金。

4. If the purchaser needs to change the mold or the part after the files which used to build the mold and mold design approval, the purchaser should pay for the additional changes, also the mold maker will free of the responsibility if the lead time delays. And the price of changes and the update lead time should be negotiated by the both parties.

译文：开模文件及模具设计图纸经买方确认后，若买方中途提出模具或零件修改，则需另付修改费，且模具制造商对修改造成的交期延迟不负责任。具体修改费用及修改时间双方再行商定。

解析：free sb/sth of/from sb/sth：使令人不快的、不需要的一类事物离开某人/某事物；使某人/某事物摆脱某事物。both parties：双方。

5. The official language of this contract is Chinese, and all the terms and interpretation should be according with the Chinese version. This contract should be submitted to Chinese law. The other details that are not listed in this contract should be kindly negotiated by the both parties.

译文：本合同正式语言为中文，所有条款及说明均遵循中文版本，且受中华人民共和国法律约束。本合同未尽事宜，双方应本着友好的态度协商解决。

解析：accord with sth：（指事物）与某事物一致或相配合，与某事物相符。

Fill in the blanks with the proper words.

1. The mold maker should design & make the molds according to the _____ or the _____ provided by the purchaser, and conform to the purchaser's requirements.

2. The mold maker should accept to be _____ money by the purchaser if the mold maker can't finish the mold in the _____ lead time. The 0.1% of the total mold price will be cut down as _____ of per day's delay.

3. The official language of this contract is Chinese, and all the terms and interpretation should be _____ with the Chinese version. This contract should be _____ to Chinese law. The other details that are not listed in this contract should be kindly _____ by the both parties.

（一）课文导读

本课介绍了模具制造合同的具体条款，包括模具制造交货期，模具设计，生产要求，付款条约及责任条约。本课在当今社会中具有实用参考价值。

（二）课文参考译文

模具制造合同

买　　方：

模具制造商：（卖方）

本份买方与模具制造商之间的模具制造合同已被双方共同认可，合同中的各项条约规定如下：

Ⅰ. 模具名称，规格，型腔和型芯用钢，以及价格如下：

模具名称	规 格	型腔和型芯用钢	价 格	备 注
		合　　计		

Ⅱ. 所有模具价格均为广州 FOB（离岸价格）。

Ⅲ. 模具制造的交货期：交货期为_____天，从模具制造商确认收到50%定金起，至模具制造商交付第一批货物止。

Ⅳ. 模具设计与制造要求：

a）模具制造商应根据买方提供的样品或图纸来设计和制造模具，并符合买方的要求。

b）水冷系统应当合适，以保证能足够冷却模具。

c）在开始制造模具之前，模具设计方案应得到买方的批准。

d）必要时，买方应提供与模具相匹配的注射机的规格和数据。

e）除了一些特殊制件，当模具正常工作时制件（产品）应能自动脱模。

Ⅴ. 付款方式：模具开始制作前，预付模具总价的 50%作为定金，模具验收合格后，付清其余50%，模具制造商确认收到所有付款后装运模具。

Ⅵ. 责任条款：

f）如模具制造商未按买方提供的样品或图纸制造模具，买方有权要求模具制造商返修或重新制造模具。

g）如模具制造商未能在指定交期内完成模具制造，则应接受买方的罚款。每延期一天将扣除模具总价的 0.1%作为罚金。

h）开模文件及模具设计图纸经买方确认后，若买方中途提出模具或零件修改，则需另付修改费，且模具制造商对修改造成的交货期延迟不负责任。具体修改费用及修改时间双方再行商定。

Ⅶ. 本合同正式语言为中文，所有条款及说明均遵循中文版本，且受中华人民共和国法律约束。合同未尽事宜，双方应本着友好的态度协商解决。

买 方：

　　地址：

　　代表：

　　（签名）_____

日期：2008_____（年____月____日）

模具制造商：

　　地址：

　　代表：

　　（签名）_____

日期：2008_____（年____月____日）

Keys to Exercises

Lesson 1
1. mold, shaping
2. magnifying
3. Die industry

Lesson 2
1. metal products, nonmetal products
2. metal processing mold, metal casting mold, clust metallurgy mold
3. press casting mold, precise casting mold

Lesson 3
1. medium
2. soft
3. such as
4. Since
5. due to

Lesson 4
1. that
2. of
3. constituent
4. but
5. subsequent

Lesson 5
1. due to
2. avoided, lead to
3. Subjecting, to
4. in addition to
5. pulls, out of

Lesson 6
1. referred to, as

2. as, as, as, as
3. able to, in
4. for, on, for
5. for, from, to

Lesson 7
1. consist of
2. on
3. from, to
4. in, in
5. for longer runs

Lesson 8
1. by, feeding, by, tripping
2. restricted to
3. identified, of, for
4. slide, into, fed, into
5. adjustable, from, to

Lesson 9
1. upon, to, as
2. out, to
3. removal, essentially
4. on, of, except in
5. Because of, from

Lesson 10
1. on, as, for
2. of, toward, in, in
3. retains, elasticity, recovery
4. out of, with, of
5. adhere to, in

Lesson 11
1. involve, take place
2. to, of, in
3. with, of, As, into, into
4. adapted, for, of
5. for, create, on, prolong

Lesson 12
1. to, on, of
2. simultaneously
3. advanced to, by, to
4. incrementally, withdrawn
5. Misalignment, excessive

Lesson 13
1. however
2. shaped
3. Because of
4. trend
5. industrial

Lesson 14
1. of
2. molecular weight
3. secondary bonds
4. because
5. but also

Lesson 15
1. that
2. Because of
3. if
4. high
5. which

Lesson 16
1. on
2. of
3. through
4. so that
5. than

Lesson 17
1. cooling
2. compression
3. thermoplastics
4. single-stage
5. aim

Lesson 18
1. gate
2. venting
3. parting line
4. cooling
5. ejected

Lesson 19
1. with
2. or
3. such as
4. at
5. because of

Lesson 20
1. of, by, by
2. necessary, filling, defect
3. excessive, on, exercised
4. sufficient, of, from
5. Liable

Lesson 21
1. introduction, was used
2. application, exclusively
3. Combined

Lesson 22
1. is fed on, is depended on
2. compared, considerably

3. versatility, cutter, subsequent

Lesson 23

1. to produce, requires
2. suited, combination
3. variation, shaped
4. consists, rotating, special

Lesson 24

1. is illustrated, coordinates
2. consists of
3. a certain

Reading Materials (13)

1. involving
2. monitoring
3. is not linked

Lesson 25

1. unified, differences
2. scientific, reasonable, formed
3. conduct, economic

Lesson 26

1. samples, drawings
2. forfeited, appointed, forfeiture
3. according, submitted, negotiated